Resisting Scientific Realism

In this book, K. Brad Wray provides a comprehensive survey of the arguments against scientific realism. In addition to presenting logical considerations that undermine the realists' inferences to the likely truth or approximate truth of our theories, he provides a thorough assessment of the evidence from the history of science. He also examines grounds for a defense of anti-realism, including an anti-realist explanation for the success of our current theories, an account of why false theories can be empirically successful, and an explanation for why we should expect radical changes of theory in the future. His arguments are supported and illustrated by cases from the history of science, including a sustained study of the Copernican Revolution and a study of the revolution in early twentieth-century chemistry, when chemists came to classify elements by their atomic number rather than by their atomic weight.

K. BRAD WRAY is Associate Professor at the Centre for Science Studies at Aarhus University in Denmark. He has published extensively on anti-realism in the philosophy of science, the social epistemology of science, and Thomas Kuhn's philosophy of science, including *Kuhn's Evolutionary Social Epistemology* (2011).

Resisting Scientific Realism

K. BRAD WRAY
Aarhus University

CAMBRIDGE
UNIVERSITY PRESS

University Printing House, Cambridge CB2 8BS, United Kingdom

One Liberty Plaza, 20th Floor, New York, NY 10006, USA

477 Williamstown Road, Port Melbourne, VIC 3207, Australia

314-321, 3rd Floor, Plot 3, Splendor Forum, Jasola District Centre, New Delhi - 110025, India

79 Anson Road, #06-04/06, Singapore 079906

Cambridge University Press is part of the University of Cambridge.

It furthers the University's mission by disseminating knowledge in the pursuit of
education, learning and research at the highest international levels of excellence.

www.cambridge.org
Information on this title: www.cambridge.org/9781108400954
DOI: 10.1017/9781108231633

© K. Brad Wray 2018

First published 2018
First paperback edition 2020

A catalogue record for this publication is available from the British Library

Library of Congress Cataloging in Publication data
Names: Wray, K. Brad, 1963– author.
Title: Resisting scientific realism / K. Brad Wray (Aarhus University).
Description: Cambridge : Cambridge University Press, 2018. |
    Includes bibliographical references and index.
Identifiers: LCCN 2018021304| ISBN 9781108415217 (hardback) |
    ISBN 9781108400954 (pbk.)
Subjects: LCSH: Science–Social aspects. | Science–History. | Astronomy–History.
Classification: LCC Q175.46 .W73 2018 | DDC 509–dc23
LC record available at https://lccn.loc.gov/2018021304

ISBN 978-1-108-41521-7 Hardback
ISBN 978-1-108-40095-4 Paperback

Contents

Figures

Acknowledgments

There are a number of people and institutions I would like to thank for the role they played in enabling me to finish this project.

First, I want to thank the many people who provided critical feedback and encouragement on various parts of the project. Specifically, I would like to thank Bas van Fraassen, Robert Westman, Kyle Stanford, Sarah Scott, Sam Schindler, Eric Scerri, Juha Saatsi, Darrell Rowbottom, Kristina Rolin, Stathis Psillos, Mark Newman, Lori Nash, Brad Monton, Moti Mizrahi, Lee McIntyre, P. D. Magnus, Tim Lyons, Peter Lewis, Larry Laudan, David Lambie, Raphaël Künstler, Chris Haufe, Todd Grantham, Michel Ghins, Greg Frost-Arnold, Don Fallis, Ludwig Fahrbach, Gerry Doppelt, Craig DeLancey, Hasok Chang, Anjan Chakravartty, Jim Brown, Sylvain Bromberger, and Roger Ariew. Parts of this project were presented at a number of conferences, including the British Society for the Philosophy of Science meetings (2010, 2016); the annual meeting of the Creighton Club in Geneva, New York (2010); the Pacific Division meetings of the American Philosophical Association (2006, 2007, 2008, 2011); the Eastern Division meeting of the American Philosophical Association (2006); and the annual meeting of the Canadian Society for the History and Philosophy of Science (2006). I also presented parts of the project at department or center colloquia at the University of Albany (2009), Aarhus University (2017), and Case Western Reserve University (2017). I would like to thank the audiences for their critical and constructive feedback.

During the 2015–2016 academic year I was on sabbatical with the intention of bringing together a first draft of this book, and I spent the fall semester as a visiting scholar in the Department of Linguistics and Philosophy at the Massachusetts Institute of Technology. Cambridge, Massachusetts, is such a pleasant environment for working on a project like this. I want to thank Sally Haslanger for being my faculty sponsor and Alex Byrne, the chair of the department, for allowing me to be a visiting scholar in the department. Three people were especially helpful to me during this time. Sylvain Bromberger read a number of chapters, and our engaging discussions led me to revisit Pierre Duhem's work with greater care. I regularly met with Lee McIntyre for lunch at the Border Café in Harvard Square to discuss our shared interests, and shared anti-realist sympathies. This was especially helpful, as he was working on his own book manuscript on a related topic. Paul Hoyningen-Huene gave a talk at the Boston Colloquium on the philosophy and history of science, and during his stay in Boston we had the opportunity to share our thoughts on the realism/anti-realism debate and philosophy of science in general. All three know how to mix philosophical discussion with laughter.

Some other individuals helped with very particular aspects of the book. I thank Eric Scerri for his patient feedback on the material on the history of chemistry. And Chris Haufe deserves a special thanks for insisting that I attempt to explain how false theories can be successful. Juha Saatsi invited me to contribute a chapter to *The Routledge Handbook of Scientific Realism*. As I prepared my contribution to the volume, I was able to catch up with a broad range of developments in the debate, which proved very helpful when I embarked on my book project. A number of experts in the history of astronomy have also willingly answered my questions. They include Robert Westman, Noel Swerdlow, Michael Shank, Owen Gingerich, and Peter Barker.

Second, I want to thank Cambridge University Press for their support. Hilary Gaskin has been a very supportive editor to work with, and has been patient throughout the process. And the two referees provided thoughtful and constructive reports on the manuscript. Having to address such astute readers has greatly improved the manuscript.

Third, the State University of New York, Oswego, granted me a sabbatical leave for the 2015–2016 academic year. I am thankful for the leave, which gave me some much-needed uninterrupted time to work on the manuscript. I am also thankful for the travel support from the Office of the Dean of the College of Liberal Arts and Sciences and the Office of International Education, which covered some of my travel costs to the various conferences at which I presented parts of the project. I would also like to thank my students in my Philosophy of Science classes at SUNY-Oswego for the opportunity to present and fine-tune some of the ideas and arguments in the book.

I continue to benefit from the ongoing support of a number of people. Paul Hoyningen-Huene, Tom Nickles, and Stathis Psillos have provided a wealth of career advice over the last seven or so years. Kristina Rolin and Hanne Andersen have been reliable sources of support for over a decade. And my partner, Lori Nash, continues to be encouraging of my projects and everyday life. She is one of my most valued critics. During the final push to get the manuscript ready, she read the entire thing in three days.

I thank the various publishers and editors for permission to publish the following papers in whole or part:

Forthcoming. "The Atomic Number Revolution in Chemistry: A Kuhnian Analysis," *Foundations of Chemistry*.

Forthcoming. "Discarded Theories: The Role of Changing Interests," *Synthese: An International Journal for Epistemology, Methodology and Philosophy of Science*. Special issue on scientific realism, guest editor: Darrell Rowbottom. DOI: 10.1007/s11229-016-1058-4

2018. "A New Twist to the No Miracles Argument for the Success of Science," *Studies in History and Philosophy of Science*, Vol. 69, pages 86–89.

2016. "Method and Continuity in Science," *Journal for General Philosophy of Science*, Vol. 47: 2, pages 363–375. DOI: 10. s10838-016-9338-8

2015. "The Methodological Defense of Realism Scrutinized," *Studies in History and Philosophy of Science*, Vol. 54, pages 74–79. DOI: 10. 1016/j.shpsa.2015.09.001

2015. "Pessimistic Inductions: Four Varieties," *International Studies in the Philosophy of Science*, Vol. 29: 1, pages 61–73. DOI: 10. 1080/02698595.2015.1071551

2013. "The Pessimistic Induction and the Exponential Growth of Science Reassessed," *Synthese: An International Journal for Epistemology, Methodology and Philosophy of Science*, Vol. 190: 18, pages 4321–4330. DOI: 10.1007/s11229-013-0276-2

2013. "Success and Truth in the Realism/Anti-realism Debate," *Synthese: An International Journal for Epistemology, Methodology and Philosophy of Science*, Vol. 190: 9, pages 1719–1729. DOI: 10.1007/ s11229-011-9931-7

2012. "Epistemic Privilege and the Success of Science," *Nous*, Vol. 46: 3, pages 375–385.

2010. "Selection and Predictive Success," *Erkenntnis: An International Journal of Analytic Philosophy*, Vol. 72: 3, pages 365–377. DOI: 10. 1007/s10670-009-9206-6

2008. "The Argument from Underconsideration as Grounds for Anti-Realism: A Defence," *International Studies in the Philosophy of Science*, Vol. 22: 3, pages 317–326.

2007. "A Selectionist Explanation for the Success *and Failures* of Science," *Erkenntnis: An International Journal of Analytic Philosophy*, Vol. 67: 1, pages 81–89. DOI: 10.1007/s10670-007-9046-1

All of these papers benefited from critical feedback from anonymous referees.

I thank Lori Nash for preparing the Index, and I thank Shaheer Husanne, Ilene Roizman, and Thomas Haynes for the care they took in the production of the book.

As of July 1, 2017, I work at the Centre for Science Studies at Aarhus University in Denmark, which is proving to be a very comfortable and congenial environment to work in. I completed the final round of revisions in Denmark.

It is a great honor to be an academic philosopher these days. It is such a great pleasure to teach university students and be able to devote time to research.

Introduction

The realism/anti-realism debate in the philosophy of science has been a hot topic since the early 1980s, with the publication of Bas van Fraassen's *The Scientific Image* and Larry Laudan's "Confutation of Convergent Realism." Inspired by these works, contemporary realists and anti-realists have principally concerned themselves with the epistemology of science. The central question in the contemporary debate has been: Do we have adequate grounds for believing that our theories are true or approximately true with respect to what they say about unobservable entities and processes? A number of philosophers have begun to feel that this particular debate has run its course, or has reached an impasse where neither side is likely ever to be in a position to claim victory.[1]

In an effort to move the debate forward, I propose to shift the focus slightly, away from the epistemic status of our current best theories to a consideration of their likely fate. I will argue that our current best theories are quite likely going to be replaced in the future by theories that make significantly different ontological assumptions. Such radical changes of theory, I argue, are irreconcilable with many forms of scientific realism. I will thus defend an anti-realist position.

The form of anti-realism that I will be defending can stand up to the significant challenges posed by realists. For example, many realists have expressed the concern that, unlike the realist, the anti-realist

---

[1] Arthur Fine (1984) has been most explicit in expressing disdain for the debate, though his concerns seem to predate the renewed enthusiasm that followed the publication of van Fraassen's book and Laudan's article. Magnus and Callender, on the other hand, have suggested that any progress in the debate can only be made at the local level, by assessing evidence for and against the existence of particular theoretical entities or posits (see Magnus and Callender 2004). They insist that the global debate is irresolvable. Rightly, Paul Dicken has recently argued that if the debate goes too local, the issues are no longer philosophical issues, but scientific issues, to be resolved by working scientists (see Dicken 2016, Chapter 5).

cannot explain the success of our current best theories. The realists' No Miracles Argument, the supposed "Ultimate Argument for Realism," builds on this conviction. The realist rightly notes that if our theories are true or approximately true, it is not at all surprising that they are successful. But the realist is not correct in claiming that the success of theories is a miracle if our theories are neither true nor approximately true. I will argue that anti-realism can offer insights into why our theories are successful, even if our theories *may* not be true or approximately true.

The structure of the book is as follows. In Part I, I take stock of the arguments against realism, addressing some common criticisms raised by realists against these anti-realist arguments. In Part II, I present some new arguments in support of anti-realism, and present a viable anti-realist explanation for the success of science. There are three key questions that a viable form of anti-realism needs to address: (i) What is the fate or expected fate of scientific theories? (ii) What warrant or epistemic support do our current best scientific theories have? (iii) How can we explain the success of science, specifically the *predictive* success of our best scientific theories, given that they may be false? My answers to these questions, in brief, are as follows.

First, the anti-realist position that I defend is a view about scientific theories and their expected fate. It is a view that is committed to the claim that radical theory change is an important part of scientific progress. In brief, a radical change of theory involves the replacement of one theory by another that carves up the world in a significantly different way, using concepts and categories that cut across the concepts and categories of the replaced theory. Such changes, I argue, pose a significant threat to a number of forms of scientific realism. There are serious challenges in reconciling such disruptive changes of theory with the realist's commitment that the concepts and categories of our current best theories cut nature at its joints, or get at the *natural groupings* of things. I grant that the key theoretical concepts and categories employed in a theory divide the things in the world into groups. But I argue that there are many different, incompatible ways of grouping things in the world, and many of these groupings could result in a predictively successful science. So a key feature of the form of anti-realism that I defend is a claim about the likely prospects of our theories. I argue that there is reason to believe that many of our best theories are apt to be rendered obsolete in the future.

Second, the anti-realist position I defend clarifies the nature and limits of the evidence in support of our scientific theories. Our best contemporary scientific theories are very impressive and capable of yielding very precise predictions of all sorts of phenomena. Our knowledge of the phenomena has been increasing markedly throughout the history of science. Some of our best theories have even led to the discovery of unexpected phenomena, phenomena that the theories were not initially designed to account for. Indeed, scientists have frequently relied on many of our best theories to predict such phenomena in advance. But there are logical considerations that set limits on the degree of warrant, justification, corroboration, or confirmation that our theories have. Consequently, I believe that a certain degree of skepticism is warranted with respect to theoretical knowledge.

Anti-realists grant that the methods of testing that scientists employ are generally very effective at identifying and weeding out *false* theories. In this, they agree with realists. But anti-realists argue that these methods are not especially well suited to determining when a theory is true, or even approximately true.[2] Our methods of testing are only applied to the various theories we have developed to date. This is a significant limitation, and greatly restricts the sorts of inferences we are warranted in drawing about our theories. Consequently, our methods are quite limited in the warrant they can provide for our theories.[3]

Third, the anti-realist position I defend argues that realists are mistaken in claiming that the best explanation for the success of science is the (alleged) fact that our theories are true or approximately true with respect to the claims they make about unobservable entities and processes. Instead, I argue that the best explanation for the success of our current best theories is the fact that unsuccessful theories have been abandoned. The methods of science enable us to determine which of the theories we have developed so far are the most successful. And scientists respond to the assessment of theories accordingly, abandoning those that do not

---

[2] This is not a new insight. Karl Popper drew attention to it in *The Logic of Scientific Discovery*, as did Pierre Duhem decades earlier (see Popper 1935/2002; Duhem 1906/1954). Oddly, though, this insight never undermined Popper's realist convictions, though his realism is quite different from most contemporary forms of realism.

[3] I also believe that anti-realists have been correct to emphasize the limited power of explanatory considerations as support for a theory. From a logical point of view, an explanation is merely a claim that is logically consistent with a theory.

measure up. But employing these methods will not necessarily lead scientists to accept a true or approximately true theory. Only if scientists have developed a true or approximately true theory would the methods of testing lead them to choose it. The methods of testing, however, are methods of justification. They are not methods of discovery.

Part I, where I focus on arguments against realism, begins with a presentation of a case study that I appeal to repeatedly in subsequent chapters. The case study concerns the history of Western astronomy, from Babylonian times to the mid-1600s, with special attention to the emergence and acceptance of the Copernican theory. Astronomy has been a well-developed science since at least 200 AD, employing mathematical models that were quite successful at generating accurate predictions. In this respect, it would count as a "mature science" by any reasonable measure. Throughout the book, I refer back to material in this chapter, as just about every important issue in the contemporary realism/anti-realism debate can be fruitfully illustrated by appealing to this case.

I then examine some logical considerations that threaten realism. I begin with a brief review of various Arguments from Underdetermination because of the central role that appeals to underdetermination have played in the debate in the past. I argue that much of the literature on the underdetermination of theory choice by evidence is irrelevant to the contemporary debate between realists and anti-realists. Most importantly, I do not believe that anti-realists should build their case on the fact that it is *logically possible* that some theory other than the currently accepted one can account for the data that our current best theory accounts for. The fact that it is logically possible that our successful theories are false gives us very little reason to be skeptical about their truth or likely truth. An anti-realism based on such an argument is not a particularly compelling position.

I then turn to consider logical considerations in support of anti-realism, most importantly the Argument from Underconsideration, alternatively referred to as the "Argument from a Bad Lot." This argument is based on the following two facts: (i) when scientists are choosing a theory, they are seldom choosing between more than a few competing theories and (ii) their evaluations of competing theories are comparative in nature. Because scientists are choosing between just a few theories, specifically the few that have been developed to date, they are not warranted in inferring that the superior theory is true or even approximately true. Moreover, *comparative evaluations* of theories are not fit to support inferences to the truth of a theory. Consequently, we do not have adequate grounds for believing that our theories are true or approximately true with respect to what they say about unobservables.

I then examine the arguments that draw on evidence from the history of science in support of anti-realism. I begin by reviewing a variety of Pessimistic Inductions, including some that have been advanced by realists in an effort to clarify the limits of our scientific knowledge. This is followed by an examination of some recent realist attempts to undermine a pessimistic inference about today's theories from a consideration of the fate of past theories. Some realists have attempted to show that today's theories are radically different from the theories of the past, and consequently are less apt to be discarded in the future than were their predecessors. These attempts, I argue, have failed. Realists have not yet identified a means to distinguish today's successful theories from the successful but now rejected theories of the past that our predecessors could not have used to distinguish their best theories from earlier successful but now rejected theories. Hence, we have little reason to think that contemporary realists have finally identified the marks of approximately true theories.

Next, I clarify the types of changes of theory that pose the greatest threat to realism. Here I draw on Thomas Kuhn's later work, in which he argues that radical theory change involves a specific sort of change to a scientific lexicon (see Kuhn 1991/2000). I illustrate this type of theory change with a detailed case study from the recent history of chemistry, when chemists came to classify chemical elements by their atomic number, thus displacing the earlier practice of classifying elements by their atomic weight. It is changes of theory of this sort that are relevant to constructing a viable Pessimistic Induction, for it is only changes of this sort that threaten Scientific Realism. Finally, I examine weaknesses in the realists' appeal to the so-called theoretical values: simplicity, breadth of scope, and such. I argue that these values do not support an inference to the likely truth or approximate truth of the theories that embody them. These values only support ordinal rankings of the various theories evaluated.

Part II, where I offer new arguments in support of anti-realism, begins by addressing a key challenge that anti-realists' face, the challenge of explaining the success of science. I first develop and defend the selectionist explanation for the success of science originally developed by van Fraassen in *The Scientific Image* (van Fraassen 1980). This explanation presents a serious challenge to the realists' No Miracles Argument, which claims that realism is the *only* philosophy of science that does not make the success of science a miracle. I then explain why false theories can generate true predictions, a challenge that many realists assume anti-realists cannot adequately address. Through a detailed examination of Ptolemy's theory and planetary models, I identify a variety of features that enable false theories and models to generate true predictions. The argument I present supplements the impressive data gathered by Tim Lyons (2002; 2006; 2012; forthcoming), Peter Vickers (2013), and others that clearly shows that many false theories have generated true predictions, even true predictions of novel phenomena.

I then spell out the details of my anti-realist position with special attention to understanding how scientists' interests influence theories and theory change. I argue that theories are always only partial representations, and scientists' interests change over time. Sometimes scientists' interests change because they have solved specific research problems. But as their research interests change, I argue, they are apt to develop new theories that make significantly different assumptions about the world, different from the assumptions made by the replaced theories.

The position I defend has much in common with van Fraassen's Constructive Empiricism, and I often appeal to his work. But my view differs from his insofar as I put more stock in the historical arguments in support of anti-realism than he does. Further, my anti-realist position builds on the work of other influential anti-realists. My view draws on Larry Laudan's critique of the so-called theoretical virtues, Kyle Stanford's New Pessimistic Induction, and Timothy Lyons's work on the predictive power of false theories. In addition, I draw on Kuhn's work on the nature of revolutionary theory change and the dynamics of normal scientific research.

Against Realism

1 | *The Copernican Revolution in Astronomy*

Many of the issues and considerations that figure in the contemporary realism/anti-realism debate in the philosophy of science can be effectively illustrated by examples from the history of astronomy. In this chapter, I want to provide a summary account of the Copernican Revolution. This will prove useful for discussions later in the book. My focus will be on considerations that are relevant to the realism/anti-realism debate.

Given that the Copernican Revolution took years to run its course, any chapter-length presentation will necessarily be selective.[1] But I will begin by providing some background about the practices of and approaches to astronomy, beginning with the Babylonians and the Greeks.

Babylonian Astronomy: The Limits of Instrumentalism

In ancient Babylon, astronomers were thoroughgoing instrumentalists. They were not concerned with cosmology, and did not even attempt to construct geometric models in their efforts to predict the phenomena. As James Evans notes, "the Babylonian planetary theory had no elaborate philosophical underpinning — there seems to have been no set of physical principles comparable to those that Aristotle provided for Greek astronomers" (Evans 1998, 22–23).

Instead, from the data they collected from observation, the Babylonian astronomers constructed tables that they used to compute predictions of noteworthy celestial events, like "the first appearance

---

[1] Ernan McMullin notes that the Copernican Revolution "took a century and a half, from Copernicus's *De Revolutionibus* to Newton's *Principia*, to consummate" (see McMullin 1993, 60). If the Copernican Revolution is understood as a revolution in astronomy, then arguably it was more or less complete by the 1630s. The Catholic Church's treatment of Galileo clearly suggests that they were on the defensive.

of the new moon" (see Evans 1998, 22; Lindberg 2007, 16). This is a
most extreme form of instrumentalism. There is *no* theory needed, no
picture of the underlying causes of the phenomena. Instead, the tables
constructed on the basis of observations provide a means for calculating when certain noteworthy or interesting celestial events, like
eclipses, conjunctions, and new moons, would occur (see Hoskin
1997a, 23–29; also Lindberg 2007, 16–17). The working assumption
was that these celestial events occurred with some sort of regularity,
and the tables were constructed to reveal the pattern in their occurrences, thus affording the astronomer the ability to predict future
occurrences. Evans notes that "part of the motivation for making the
observations was religious. And part of it was practical: the stars and
especially the planets were believed to provide signs of the future
welfare of the king and the nation" (Evans 1998, 14). But these goals,
at least as they were understood by Babylonian astronomers, could
be effectively realized without venturing into cosmology.

Astronomy in Ancient Greece: Two Traditions

The ancient Greeks approached astronomy in a different manner and
with different mathematical tools than the Babylonians. The Greeks
used geometry, and attempted to construct geometric models. These
models were intended to reflect at least some of the features of the
structure of the cosmos. For example, the models were more or less
Earth-centered, on the assumption that the Earth was at the center of
the cosmos (see Evans 1998, 76).

There were two somewhat distinct research traditions in ancient
Greek astronomy, one principally concerned with cosmology and the
other principally concerned with prediction. The tradition concerned
principally with cosmology sought to model the planets' orbits using
nested orbs, or homocentric spheres (see Hoskin 1997a, 34). Eudoxus
of Cnidus, a near contemporary of Plato, developed such models in an
effort to account for retrograde motion, the apparent backward motion
that the planets periodically go through in their cycles. According to
Michael Hoskin,

The astronomer was to imagine the planet located at the equator of a
sphere that was spinning uniformly. Projections were protruding from
these poles, and these were embedded into a second sphere, outside the

first and concentric with it. This outer sphere was also spinning but about a somewhat different axis, and as it spun it carried with it the inner sphere. In consequence, the movement of the planet reflected the spinning of both spheres. Eudoxus realized that if the two spins were equal in speed but in opposite direction, and if the two axes were not very different, then the planet would move back and forth in a figure-eight. (Hoskin 1997a, 34–35)

Eudoxus introduced two additional spheres for the models of each of the planets Mercury, Venus, Mars, Jupiter, and Saturn. The combined motion of the four spheres for each planet could, at least in principle, account for the motion of the planet, both its periodic retrograde motion and its daily motion in the opposite direction of the fixed stars (see Hoskin 1997a, 35). These planetary models that employed homo-centric spheres were three-dimensional and were thought to provide a physical representation of the cosmos. But such models were never developed in sufficient detail or with adequate precision to enable astronomers to make accurate predictions.

The second research tradition in the ancient Greek world was prin-cipally concerned with developing models that could generate accurate predictions. Astronomers showed great ingenuity in the mathematical models they developed for this purpose. Hipparchus, for example, introduced the eccentric circle in his model of the orbit of the Sun (see Evans 1998, 211). With an eccentric circle, the Earth is placed not at the center of the Sun's orbit, but rather off-center (see Figure 1a). Hipparchus' eccentric circle model of the Sun's orbit enabled him to account for the varied lengths of the seasons.

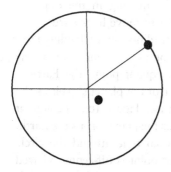

Figure 1a Eccentric circle model
The Earth is off-center from the point that is the center of the planet's motion.

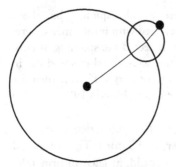

Figure 1b Epicycle and deferent circle model
The planet travels around the epicycle as the epicycle is carried around the deferent circle.

The eccentric circle was just one of a number of mathematical devices introduced by Greek astronomers in their efforts to account for the motion of the planets and predict their locations accurately. Even before Hipparchus, Appolonius introduced the epicycle and deferent circle model (see Evans 1998, 22). With this type of model, a planet moves around a circle, the epicycle, which has its center placed on another circle, the deferent circle, which orbits the Earth (see Figure 1b). The epicycle and deferent circle model fulfilled two functions. First, using such a model, astronomers were able to make more accurate predictions of the various planets' locations. Second, the epicycle and deferent model could account for the apparent backward motion of the planets as they moved through retrograde motion. When the planet is on the inside of its epicycle, moving in the opposite direction of its deferent circle, it will appear to move backward.

The culmination of this model-building tradition was Claudius Ptolemy's *Almagest*. Ptolemy is responsible for another innovation, the equant point. With models employing an equant point, the Earth is placed at the same distance from the center of the planet's orbit as the equant point, but in the opposite direction (see Figure 1c). The equant point is the center of motion for the planet as the planet orbits the Earth. The planet sweeps out equal angles in equal time around the circle describing its orbit. But because the equant point is off-center, viewed from the Earth, the planet will appear to move at different speeds through its orbit. By employing combinations of equant points, epicycles

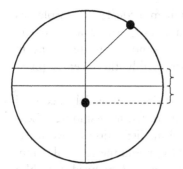

Figure 1c Equant point model
The center of the planet's motion is placed a distance from the center of the circle equal to the distance the Earth is placed, but in the opposite direction.

and deferent circles, and eccentric circles, Ptolemy created the most accurate mathematical models that had ever been developed. In fact, they were not surpassed in accuracy for centuries. Even Copernicus' mathematical models were no more accurate than Ptolemy's.

This approach to model-building is clearly in the instrumentalist tradition. The models were two-dimensional rather than three-dimensional, like the models employing homocentric spheres. And *generally* it was assumed that these two-dimensional models did not describe the structure of the cosmos. Some of the models employed devices that were widely regarded as difficult to reconcile with a physical picture of the cosmos. Rather, the value of these models was their ability to generate accurate predictions.

It is worth drawing a distinction between the following activities that ancient Greek astronomers engaged in: (i) observational astronomy, (ii) mathematical astronomy, (iii) cosmology, and (iv) astrology. Many of the people we identify as astronomers from this period were involved in more than one of these enterprises. Still, it is worth distinguishing the various activities from one another. The distinction between mathematical astronomy and cosmology is especially relevant to our concerns. Insofar as one worked in mathematical astronomy, one was generally an anti-realist, and often an instrumentalist. One did not need to assume that a planet's orbit was as complex in structure as the geometric model of its orbit. The aim of these models was to save the phenomena, that is, to account for the observables, and to predict future observable events (see Duhem 1908/1969). Insofar as one worked in cosmology, one was a realist. Many astronomers aspired to represent the structure of the cosmos accurately. And in this endeavor, the astronomer was constrained by the accepted physical theories.

Aristotle provided the most comprehensive and widely accepted physical theory for Greek astronomers interested in cosmology. Central to Aristotle's physics was a categorical distinction between the terrestrial realm, all that is beneath the Moon, and the celestial realm, which included the Moon, the Sun, the planets, and the fixed stars. Not only did these two realms operate according to different principles, they were made of fundamentally different substances. Whereas everything in the terrestrial realm was thought to be made of a combination of earth, water, air, and fire, everything in the celestial realm was thought to be made of ether or quintessence, an indestructible element. The natural motion of things in the terrestrial realm was either upward toward the heavens or downward toward the center of the Earth, depending on the constitution of the particular thing in question. The natural motion of things in the celestial realm was circular, a fitting motion for indestructible things, as it was the only motion that was eternal.

Medieval and Renaissance Astronomy

Astronomy was one of the casualties of the breakdown of the Roman Empire and the barbarian invasions in Europe. Even just a century before Copernicus was born, European astronomers still had a far less sophisticated understanding of astronomy than Ptolemy and his contemporaries (see Kuhn 1957, 124). But by the Renaissance, the knowledge of Greek astronomy was almost fully retrieved, partly with the aid of texts from the Islamic world.[2] Europeans finally reached the

---

[2] In the last few decades, historians have reassessed the influence of Islamic astronomy on developments in European astronomy. Some of the important developments made in the Islamic world may have been independently discovered later by Europeans (see, for example, Gingerich 1974/1993, 175). The model that Copernicus used to eliminate Ptolemy's equant point, for example, "was precisely the same mechanism suggested two centuries earlier by Ibn ash-Shātir in Damascus" (Gingerich 1974/1993, 175). But some developments were adopted from Islamic sources, for example, a particular device developed by Nasīr al-Dīn al Tūsī, the Tūsī device, which enabled astronomers to model rectilinear motion

same level of sophistication with respect to mathematics that Ptolemy and his contemporaries had. Crucial in this process was the publication of *Epitome of the Almagest* by Georg Peuerbach and his student Johannes Regiomontanus (see Lindberg 2007, 162).[3]

Regiomontanus was a committed realist. He objected to the planetary models in Ptolemy's *Almagest*, partly on the grounds that they were two-dimensional, and thus incapable of accurately representing the causes of the planets' motions. Regiomontanus insisted that three-dimensional models, models employing homocentric spheres, were required in order to accurately account for the causes of planetary motion (see Shank 2002, 186). Regiomontanus had hoped to develop models employing three-dimensional homocentric spheres that would both account for the physical causes of the planets' motions and enable astronomers to derive accurate predictions of their locations (see Shank 2002, 192). Thus, Regiomontanus "contradicts the late-antiquity interpretation of Aristotle's distinction between natural philosophy, on the one hand, and mathematics and astronomy, on the other" (Shank 2002, 192). Even though Regiomontanus never managed to realize his goal, his research was driven by a firm commitment to realism.

Interestingly, Regiomontanus also drew attention to the fact that one could model the motions of Saturn, Jupiter, Mars, Venus, and Mercury using either "a deferent carrying an epicycle that rotates about its center with the speed of the mean Sun" or "an eccentric whose center is moving about the Earth with the speed of (and in the same direction as) the mean Sun" (see Shank 2002, 183–184).[4] Both models were capable of generating predictions that were equally accurate.[5] This is a

---

using circles (see di Bono 1995). Copernicus employed such a device in some of his planetary models, though, as Mario di Bono explains, it is unclear exactly how he acquired knowledge of Tūsī's work (see di Bono 1995, § 4 and § 7).

[3] Noel Swerdlow argues that Regiomontanus "was the *only* person of his age to understand astronomy well enough to single out its faults" (Swerdlow 2004, 85).

[4] A remark is in order about the notion of the "mean Sun." "The *mean Sun* is a fictitious body that moves uniformly on a circle centered at the Earth" (see Evans 1998, 226). "The mean Sun lies in the same direction as the true [Sun] whenever the true Sun is in the apogee … or the perigee … of its eccentric circle … At all other times of the year, the true Sun, as seen from Earth, is a little ahead of or a little behind the mean Sun" (226). A number of the constraints in Ptolemy's planetary models are, strictly speaking, related to the mean Sun rather than the true Sun. But for our purposes, such details need not concern us.

[5] Ptolemy had already noted that the orbits of the superior planets, Mars, Saturn and Jupiter, could be modeled with either eccentric circles or epicycles, but it

vivid example of the underdetermination of theory choice by evidence. Given the evidence available, astronomers were unable to determine which model was closer to the truth. Such underdetermination has often been appealed to by anti-realists as evidence that scientists should be cautious about inferring that a theory is true or even approximately true on the basis of the fact that it can account for the phenomena and generate accurate predictions. I discuss arguments from underdetermination in detail in the next chapter. Though the underdetermination of theory choice by evidence raises challenges for scientific realism, I do not believe that it is the strongest argument in support of anti-realism.

Copernicus: The New Cosmology

Copernicus was fortunate to be born at a time when the knowledge of the mathematics necessary to understand and appreciate Ptolemy's accomplishments was restored. According to David Lindberg, Peuerbach's and Regiomontanus' *Epitome* "exercised a strong influence on Nicolas Copernicus" (Lindberg 2007, 162). But dissatisfied with the two traditions in ancient Greek astronomy, one based on homocentric spheres and the other employing epicycles and deferent circles, Copernicus was ultimately led to develop a radical new theory, a heliocentric theory of the cosmos (Copernicus 1543/1995, 5).

It is worth stressing how radical Copernicus' cosmology was. With the Copernican Revolution in astronomy, the kind-term "planet" changed its meaning. According to the lexicon of the Ptolemaic theory, the term "planet" picked out the following entities: the Moon, Mercury, Venus, the Sun, Mars, Jupiter, and Saturn. With Copernicus' new theory, some of the things previously regarded as planets were no longer regarded as planets. Specifically, the Sun and the Moon were no longer regarded as planets. And the Earth, which was previously not regarded as a planet, was thenceforth regarded as a planet. Not only are there changes in the extensions of key theoretical concepts, but their intensions change as well. In Ptolemy's theory, "planet" meant "wandering star." The planets were just those "stars" that did not move with the fixed stars. Each of the Ptolemaic planets had its own motion, generally in the direction opposite to the motion of the fixed

appears that he did not realize that the orbits of the inferior planets could also be modeled either way (see Shank 2002, 183).

stars, in addition to the daily motion it shared with the fixed stars. In Copernicus' theory, "planet" meant "a satellite of the sun," and the fixed stars were truly fixed. Their apparent daily motion was just that, *apparent* motion, the result of the fact that the Earth completes a rotation on its axis each day.

Copernicus was a realist about his radical new cosmology. In arguing for the truth of his new cosmology, he appealed to various theoretical virtues. He argued that his theory captured the simplicity and harmony of the cosmos better than Ptolemy's theory (see Copernicus 1543/1995, 24 and 26; Gingerich 1975b/1993, 199). Copernicus identified a number of facts that his theory could explain, including the following:

(1) Saturn, Jupiter, and Mars are in opposition to the Sun when "they are nearer to the Earth";
(2) Mars appears so much larger when it is in opposition to the Sun; and
(3) Mercury and Venus never stray far from the Sun (see Copernicus 1543/1995, 27; and 21–22).[6]

These facts are to be expected, given the order of the planets and the structure of the cosmos in the Copernican theory. Though these facts were recognized by Ptolemy, he had accounted for them in ad hoc ways. For example, in order to account for the constrained orbits of Mercury and Venus, Ptolemy just stipulated that the center of the epicycle of each of these planets always remained on a line running from the center of the Earth to the Sun. Copernicus, on the other hand, argued that the reason that Mercury and Venus never appeared far from the Sun was because their orbits were contained within the orbit of the Earth. And Ptolemy accounted for the fact that Mars appeared larger when it was in opposition to the Sun by ensuring that Mars was on the inside of its epicycle when it was in opposition. Copernicus, on the other hand, argued that Mars appeared larger when it was in

---

[6] Swerdlow identifies a number of other harmonies that Copernicus could explain, given his heliocentric theory (see Swerdlow 2004, 88–90). Importantly, Swerdlow even identifies some erroneous relations that followed from Copernicus' theory, which Copernicus took to be evidence for his theory (see Swerdlow 2004, 90). This is a recurring theme in the history of science, and one relevant to the realism/anti-realism debate. Scientists are as committed to the false claims entailed by their theories as they are to the claims we still regard as true.

opposition to the Sun because it was much closer to the Earth then than when it was in conjunction with the Sun.

Realists often appeal to theoretical values such as simplicity and harmony as evidence that a theory is approximately true. Many realists insist that these values are reliable indicators of the truth or approximate truth of our theories. Indeed, Copernicus seems to have reasoned this way (see McMullin 1993, 72–74). He took the simplicity and harmony of his theory to be indicators of its likely truth.

Copernicus' theory, though, did face some serious challenges. In order to develop a theory that was as accurate at predicting the locations of planets as the contemporary Ptolemaic theory was, Copernicus had to employ both (i) eccentric circles and (ii) epicycles and deferent circles in his planetary models, irregularities that he regarded as misrepresenting the structure of the cosmos. Hence, even granting that the Copernican theory is a more accurate representation of reality insofar as it acknowledges that the Earth and other planets orbit the Sun, its predictive success was not a consequence of the fact that it mirrored reality. Rather, the theory's predictive success was a consequence of the fact that it employed eccentric circles, epicycles, and deferent circles. These were built into the planetary models, ad hoc, to ensure that the theory could account for the phenomena, and that it would be as successful as the contemporary Ptolemaic theory.[7]

Even with these ad hoc adjustments, Copernicus' planetary models were no more accurate than the late Renaissance version of the Ptolemaic models of planetary motion.[8] In fact, the two theories erred by as much as 5 degrees with respect to some predictions (see Thoren 1967; Gingerich 1975b/1993, 195–196; 1971/1993). Generally, though, both theories were impressively accurate. For example, Owen Gingerich notes that the predictions for the Moon derived from the Prutenic Tables, which were based on Copernicus' models, erred on average only 30 minutes of an arc, that is, half a degree. And the predictions for

[7] Unlike Ptolemy, Copernicus used only minor epicycles in his planetary models. Major epicycles were introduced to model retrograde motion. Minor epicycles have a different function. They modify the shape of a planet's orbit in an effort to achieve greater predictive accuracy, but they do not account for retrograde motion (see figure 22 in Kuhn 1957, 67).

[8] The version of the Ptolemaic theory popular in Copernicus' day was not much different from Ptolemy's original theory (see Swerdlow 2004, 80). It was not radically modified in the intervening years, contrary to the popular myth that suggests otherwise (see Swerdlow 2004, 79–80; Gingerich 2004, chapter 4).

the superior planets, Mars, Jupiter, and Saturn, "tended to agree within 10 [minutes of arc]" (see Gingerich 1978/1993, 210).

Despite the fact that Copernicus still needed to employ epicycles in his models, he did offer astronomers something of interest. Copernicus' mathematical models of the planets did not require the use of equant points. Some astronomers had found the equant point especially objectionable. An equant shifts the center of motion away from the center of the circle that defines the orbit of the planet. And the planet sweeps out equal angles around the circle in equal time. Because the Earth is not situated at the equant point in Ptolemy's models, a planet will sometimes appear to move faster through the stars, and sometimes slower. Copernicus and other astronomers regarded the equant as a violation of one of the first principles of mathematical astronomy, that "celestial motion is circular and uniform, or composed of circular and uniform parts," a principle that allegedly originated with Plato (see Gingerich 2004, 53; also Duhem 1908/1969, 5). Copernicus insisted that because the irregular motions of the planets, that is, their periodic retrograde motions, had "fixed periodic returns," the movements of the planets must be "circular or composed of many circular movements" (Copernicus 1543/1995, 12). Only circular motions, he thought, could ensure the regularity in the pattern of their phases of retrograde motion.

Osiander: The First Instrumentalist Reading of Copernicus

Georg Joachim Rheticus carried Copernicus' manuscript from Poland to Nuremberg to deliver it to the printer (see Gingerich 1974/1993, 167). But Andreas Osiander was left with the responsibility for seeing the book through its publication. As a consequence of this role, Osiander is responsible for the first anti-realist reading of Copernicus' theory.

Osiander attached an unsolicited and unsigned introduction to Copernicus' manuscript, titled "To the Reader Concerning the Hypotheses of this Work." In the introduction, Osiander urges the reader to not assume that the hypotheses about the motions of the planets presented in the book describe the real motions of the planets. So rather than taking Copernicus' models to be accurate descriptions of the structure of the cosmos, Osiander suggests that they should be taken merely as useful means for determining the locations of the planets (see Osiander in Copernicus 1543/1995, 3–4). Importantly,

Osiander failed to make it clear to the reader that it was he, and not
Copernicus, who prepared the introduction. Thus, many early readers
of the book thought that Copernicus was an instrumentalist. There is
some debate among contemporary historians of science about whether
Osiander is best characterized as an instrumentalist or a fictionalist.
Instrumentalists regard a theory as a mere instrument for prediction
and control, and not the sort of thing that one should regard as either
true or false. Fictionalists, on the other hand, do not believe that the
world is as a theory describes it. But they urge us to act "as if" the
world were as the theory suggests (on fictionalism, see Suárez 2009).
Alternatively, some argue that Osiander thought that knowledge of the
underlying structure of the cosmos exceeded our human capabilities
(see, for example, Barker and Goldstein 1998; Shank 2002). The latter
view was attributed to Osiander by Edward Rosen. In Rosen's words,
Osiander believed that "since divine revelation is the only source of
truth, astronomical hypotheses are not concerned therewith, and serve
only as a basis of calculations" (see Rosen 1939/1959, 25).[9] In the
contemporary realism/anti-realism debate, this latter position is most
in line with Bas van Fraassen's constructive empiricism, though van
Fraassen does not claim that divine revelation is the only source of
truth (see van Fraassen 1980).

Though Osiander's motives for attaching the introduction to Copernicus' book are unclear, it is interesting to note his reasoning as stated
there. Osiander claims that the discipline of astronomy "is absolutely
and profoundly ignorant of the *causes* of the apparent irregular movements" of the planets (Osiander in Copernicus 1543/1995, 3; emphasis
added). Thus, he warns readers that

as far as hypotheses go, let no one expect anything in the way of certainty
from astronomy, since astronomy can offer us nothing certain lest, if anyone
take as true that which has been constructed for another use, he go away
from this discipline a bigger fool than when he came to it. (Osiander in
Copernicus 1543/1995, 4)

---

[9] When Copernicus' book was published, the Reformation was in full force and the
people involved in astronomy were also deeply engaged with the religious
changes that were sweeping across Europe. For example, Osiander was raised a
Catholic, and even taught at an Augustinian cloister in Nürnberg. But he later
became a Lutheran and "gained ... notoriety as an articulate, zealous reformer
and a militant anti-Romanist" (see Wrightsman 1975, 218).

Copernicus died almost immediately upon receiving a printed copy of his book. Consequently, he was unable to correct matters and ensure that readers knew that Osiander's introduction did not reflect his own views. Rheticus knew who wrote the introduction, and he was not pleased with it at all (see Gingerich 1974/1993, 167). But it was only in 1609 that it became widely known who wrote the introduction, when Johannes Kepler sought to correct the record on this matter, motivated in part by his own realist convictions for a heliocentric theory (see Dreyer 1906/1953, 321; Duhem 1908/1969, 68–69; Koestler 1959/ 1964, 169–175). This, however, was many decades after Copernicus died.

Incidentally, as Pierre Duhem notes, Osiander had written a letter to Copernicus on April 20, 1541, two years before Copernicus' book was published (see Duhem 1908/1969, 68). In the letter, Osiander claims that

as for hypotheses, this is what I have always thought on that subject: they are not articles of faith, they are merely the basis of calculation; even if they should be false, that hardly matters, so long as they reproduce the φαινόμενα of the movements exactly. For consider, if we follow Ptolemy's hypotheses, who can assure us whether the irregular movement of the sun occurs rather in virtue of the epicycle or in virtue of the eccentric, since it can be produced in either way? I would urge you to touch on this question in your preface; you would thereby pacify the Peripatetics and theologians whose opposition you fear. (Osiander cited in Duhem 1908/1969, 68)

Kepler apparently took this as evidence that Osiander's preface did not even represent Osiander's own thoughts on the issue. Duhem contests this interpretation (see Duhem 1908/1969, 68–69; see also Rosen 1939/1959).

Reinhold and the Wittenberg Astronomers

The initial response to Copernicus' theory was rather tepid. In fact, Robert Westman suggests that by 1600, over fifty years after the publication of Copernicus' book, only ten astronomers accepted Copernicus' theory as a *true description of the world* (Westman 1986/2003, 54).[10] Aside from Rheticus, there were no astronomers

---

[10] Westman claims that "we can identify only ten Copernicans between 1543 and 1600: ... four were German (Rheticus, Michael Maestlin, Christopher Rothmann, and Johannes Kepler); the Italians and English contributed two each (Galileo and Giordano Bruno; Thomas Digges and Thomas Harriot); and the

committed to a realist interpretation of the Copernican theory until the 1570s (see Westman 2011, 148). Before 1570, many of the astronomers who objected to Copernicus' theory did so because it failed to fit with the accepted physical theory, Aristotle's theory. This concern is clearly a realist concern.

But Copernicus' theory was adopted in the 1550s and 1560s, at least in part, by the Wittenberg astronomers. Westman notes that "the principal tenet of the Wittenberg viewpoint was that the new theory could only be trusted within the domain where it made predictions about the angular position of a planet" (Westman 1975, 166). The Wittenberg astronomers believed that "the least satisfactory Copernican claim was the assertion that the earth moved" (see Westman 1975, 167; Barker 2001). Thus, though the Wittenberg astronomers enthusiastically attended to Copernicus' work, they adopted an anti-realist – specifically an instrumentalist – stance toward his theory, regarding the planetary models as merely useful devices for predicting the positions of the planets. This instrumentalist version of anti-realism was quite popular among sixteenth-century European astronomers (see Westman 2011, chapter 5). The Wittenberg astronomers were impressed with the Copernican planetary models, in part because Copernicus was able to dispense with equant points, Ptolemy's innovation.

The publication in 1551 of Erasmus Reinhold's Prutenic Tables played a crucial role in the development of the Wittenberg School (see Westman 2011, 141). The Prutenic Tables enabled astronomers to calculate and thus predict the locations of planets, and they were explicitly based on Copernican models. But Reinhold, a professor at the University of Wittenberg, emphatically rejected Copernicus' new cosmology (see Duhem 1908/1969, 73–74).

It is worth contrasting the Wittenberg astronomers' views with the view of a contemporary proponent of the Ptolemaic theory. Christopher Clavius, one of the most influential Ptolemaic astronomers of the sixteenth century, was a realist about Ptolemy's *geocentric* cosmology. Unlike Copernicus and the Wittenberg astronomers, Clavius was also a realist about eccentrics and epicycles, noting that "since up to now no one has found a more convenient method than the one that saves the

Spaniards and Dutch but one each (Diego de Zuniga; Simon Stevin)" (Westman 1986/2003, 54).

appearances by means of eccentrics and epicycles, it stands to reason that the celestial spheres have orbits of this kind" (Clavius 1581, cited in Duhem 1908/1969, 94; see also Lattis 1994, 110 and 129). Clavius' realist attitude about the eccentrics and epicycles in the Ptolemaic planetary models may not have been the standard view among late sixteenth-century Ptolemaic astronomers. But he was probably not the only one to hold this view.[11] Clavius' line of reasoning is an example of an Inference to the Best Explanation, a pattern of reasoning often employed by realists and equally as often criticized by anti-realists.

Tycho Brahe: Advances in Observational Astronomy

By the late 1580s, astronomers had even more theories to choose between. Tycho Brahe developed a new theory of the cosmos. Like Copernicus, Brahe also regarded "the equant as an abomination" (Thoren 1990, 91). But Brahe was not prepared to accept a theory that was contrary to the accepted physics. He did not believe that the Earth moved, as Copernicus claimed.

According to Brahe's theory, the Earth is at the center of the cosmos. The Moon and the Sun orbit the Earth, but the remaining planets, Mercury, Venus, Mars, Jupiter, and Saturn, orbit the Sun (see Brahe 1588/1970, 58–66). The planets are thus swept around the Earth each day with the Sun, as each planet simultaneously completes its own orbit around the Sun – Saturn in about thirty years, Jupiter in about twelve years, Mars in about two years, etc.

Importantly, though Brahe accepted Aristotelian physics, he did not slavishly follow Aristotle. On the basis of his own observations in 1577 and 1585, he came to believe that comets were not restricted to the terrestrial realm, that is, between the Moon and the Earth (see Thoren 1990, 123 and 265). In fact, Brahe determined that the comets he observed crossed the paths of multiple planets. This led him to reject the existence of the celestial spheres, allegedly made of quintessence or ether.

With the development of Brahe's theory, early modern European astronomers had three well-developed theories to choose from – Ptolemy's theory, Copernicus' theory, and Brahe's theory – and each

---

[11] Apparently, even some loyal followers of the Ptolemaic theory borrowed from Copernicus. Specifically, Conrad Dasypodius and Caspar Peucer reworked "Ptolemy's theories utilizing the constants determined by Copernicus" (see Thoren 1990, 91).

was as empirically successful as the others. That is, the three theories were equally accurate with respect to their predictions. Importantly, the three principal theories that astronomers considered during the late sixteenth and early seventeenth centuries are not *empirically equivalent* theories. They do, after all, entail radically different predictions about the world. But given the data then available, astronomers could not determine unequivocally which theory was superior. This is a case of what Lawrence Sklar calls *transient* underdetermination (Sklar 1975; see also Stanford 2001). I will return to a discussion of transient underdetermination and its relevance to the contemporary realism/ anti-realism debate in the next chapter.

Significantly, Brahe's theory had an important influence on the Copernican Revolution in astronomy, even though it never enjoyed the status as *the* accepted theory. Confronted with three plausible theories, astronomers could juxtapose each theory with the other two competitors. Such comparisons enabled astronomers to scrutinize each of the competing theories more thoroughly (on the value of comparative evaluation, see Feyerabend 1988).

Brahe's contribution to the Copernican Revolution is not limited to his new theory and the effects it had on the debates. He is also responsible for raising the standards in observational astronomy. He had numerous astronomical instruments custom built that far surpassed in accuracy any instruments that had been used before (see Brahe 1598/ 1946, 65 and 79; also Gade 1947, 84 and 85). None of Brahe's instruments were capable of magnifying, but their size and craftsmanship made it possible to achieve degrees of accuracy in observational astronomy hitherto unachieved (see Thoren 1990, 190–191). And he employed a team of astronomers, both to gather the data and to assist with processing it (Brahe 1598/1946, 67, 70, and 74; Christianson 2000, 80). At his observatory in Denmark, his team of astronomers gathered data systematically and regularly, making about "85 observing sessions a year" (see Thoren 1990, 201 and 220). Brahe had different teams take observations of the same stars and planets from different locations on his estate in an effort to detect errors (Gade 1947, 69 and 90). Consequently, he was able to gather an extensive body of data on the locations of the stars and planets that far surpassed in accuracy and scope what astronomers had been used to working with. Brahe's data would eventually play a crucial role in Kepler's important contributions to the Copernican Revolution. Prior to Brahe's innovations in observational astronomy, astronomers tended to rely on just a few data points in their efforts to model the orbits of the planets.[12]

Kepler and Galileo

Kepler and Galileo Galilei, two of the early converts to the Copernican theory, were realists about the Copernican cosmology. Importantly, though, the "Copernican" theory that Kepler and Galileo accepted departed significantly from Copernicus' own theory. Both Galileo and Kepler rejected the existence of the celestial spheres that were once thought to carry the planets in their orbits. Galileo did not think that the moon was made of ether or quintessence, as Copernicus had believed. In fact, Galileo's telescopic observations of the Moon, with its mountainous terrain, provided compelling evidence that the Moon was made of the same substance as the Earth. And Kepler believed that the planets moved in ellipses, not circles or combinations of circles.[13]

Kepler's chief contribution to the Copernican Revolution was his discovery of his famous laws of planetary motion, especially the first two laws: (i) the orbits of planets are ellipses, with the Sun located at one focus, and (ii) the planets sweep out equal areas in equal times as they move around the Sun (see Dreyer 1906/1953, 392). It is easy to exaggerate Kepler's significance in the Copernican Revolution. In *The New Astronomy* (*Astronomia Nova*), where Kepler first published the two laws, he only shows that they apply to Mars. A decade later, when he published *Epitome of Copernican Astronomy* (*Epitome Astronomia Copernicae*), Kepler just assumes that the laws apply to the other planets as well (see Kepler 1618–1621/1995; Dreyer 1906/1953, 403). Further, the discoveries for which Kepler is rightly famous were buried in texts that included many more claims that, to modern readers, would sound truly bizarre. For example, his claim that there could be only six planets was

---

[12] Compare, for example, Gingerich's (1971/1993, 379–380) discussion of Ptolemy's methods for determining the orbit of Mercury with his discussion of Kepler's resources as a result of Brahe's vast store of observations (see 1975a/1993, 340).

[13] Other early adopters of the Copernican theory also held views that departed significantly from Copernicus' own view on key points. Thomas Digges and Giordano Bruno, for example, believed that the universe was infinite in size, despite the fact that this was not Copernicus' own view. And, like Kepler, Bruno believed that there were no celestial spheres carrying the planets (see Tredwell and Barker 2004).

derived from his neo-Platonist convictions. Because there are only five Platonic solids, Kepler argued that there could be only six planets, each separated from the next by one of the five Platonic solids. Kepler also claimed that the planets were moved by a magnetic force emanating from the Sun as it turned on its axis (see Dreyer 1906/1953, 394–398). Importantly, Kepler's realism was as much tied to these claims that are now widely regarded as false as it was to the claims we continue to accept today.[14]

Further, Kepler's laws were not immediately accepted by other astronomers, not even by Galileo, despite the fact that Galileo and Kepler had corresponded on a number of occasions (see Shea and Artigas 2003, 26). Galileo's *Dialogue on Two Chief World Systems* makes no mention of elliptical orbits. Certainly by Newton's time the significance of Kepler's laws was recognized and appreciated. But by then the revolution in *astronomy* was over.

Galileo's contributions to the Copernican Revolution were more far-ranging and had a more immediate impact than Kepler's contributions. Galileo is most famous for his contributions to observational astronomy, as he was the first to employ the telescope as an instrument in astronomy. Three of his telescopic discoveries played a critical role in the Copernican Revolution.

First, his discovery that the Moon has an uneven surface, with mountains and valleys, not unlike the Earth's surface, helped erode the traditional celestial/terrestrial distinction that seemed to fit so well with the Ptolemaic theory (see Galilei 1610/2008, 51–63). After Galileo reported his observations of the Moon, it seemed untenable to insist that it was made of quintessence, a perfect and immutable substance, given that the Moon's surface so closely resembled the Earth's surface.[15] Second, his discovery of the Medicean stars, that is, the moons of Jupiter, diminished the importance of a key criticism against the Copernican theory (see Galilei 1610/2008, 68–84). According to the

---

[14] Dreyer maintains that "there is ... the most intimate connection between [Kepler's] speculations and his great achievements; without the former we should never have had the latter" (see Dreyer 1906/1953, 410). Dreyer thus suggests that Kepler's speculative metaphysical ideas, even the erroneous ones, may have played a constructive role in leading him to make the significant and lasting discoveries for which he is rightly famous.

[15] Galileo's observations of sunspots were also a challenge for the view that the cosmos was immutable.

Copernican theory, the Moon orbits the Earth as the Earth orbits the Sun. This struck many critics as impossible. They found it incomprehensible that the Moon would be able to keep up with a moving Earth as the Earth orbited the sun. The Ptolemaic theory faced no such problem, as it posits that all the celestial bodies orbit the Earth, which is stationary in the center of the cosmos. But with the discovery of the moons of Jupiter, even Ptolemaic astronomers needed to explain how a satellite stays in orbit around an orbiting planet.

Third, Galileo's discovery that Venus exhibits the full range of phases in the course of its orbit cast considerable doubt on the Ptolemaic theory. Even before making the requisite observations, Galileo predicted that Venus would exhibit a full range of phases, as the Moon does. This prediction was derived from the Copernican theory.[16] This is a classic case of deriving a prediction of *novel* phenomena from a theory and testing the theory against the world. After months of observation, Galileo's prediction was vindicated. This was the most damaging of the telescopic discoveries for the Ptolemaic theory. In constructing a model for the orbit of Venus, Ptolemy stipulated that the center of its epicycle lies on a line running from the center of the Sun to the Earth. Because of this ad hoc stipulation, the Ptolemaic theory predicted that Venus would not exhibit a full range of phases (see Galilei 1615/2008, 127). This particular discovery marked a significant turning point for the Ptolemaic theory. It was impossible to reconcile it with the observations of Venus's phases.

Galileo's contributions to the Copernican Revolution extend beyond astronomy, narrowly construed. Galileo also conducted important research in physics and hydrostatics that ultimately proved relevant to the revolution in astronomy. His work on falling bodies and floating bodies challenged the physics that was alleged to support the Ptolemaic theory. This was a crucial development in the Copernican Revolution, as one of the key points of resistance against the Copernican theory was that it conflicted with Aristotle's physics. Galileo was giving astronomers and natural scientists reason to believe that Aristotle's physics was inadequate.

---

[16] Galileo sent a message to Kepler, encoded in an anagram, announcing his prediction, partly as a means to secure his priority (see Swerdlow 1998, 260; Shea 1998, 221–222).

Arguably, by the second decade of the 1600s, the Ptolemaic theory was no longer regarded as a serious contender (see Hoskin 1997b, 130–131). Galileo's telescopic observations, especially the phases of Venus, contributed significantly to undermining astronomers' allegiance to the Ptolemaic theory. But it took longer for the battle between the Tychonic theory and the Copernican theory to be resolved, as Galileo's observations were compatible with both theories.

Like Kepler, though, Galileo was mistaken about some matters. For example, he believed that his strongest argument in support of a moving Earth was his argument from the tides. In the *Dialogue Concerning the Two Chief World Systems*, Galileo argues that the tides are caused by the combined daily motion of the Earth and the Earth's annual motion around the Sun. In fact, he insists that the tides would be inexplicable were the Earth stable and not moving, as Ptolemy maintained (see Galilei 1632/2001, 484). He reasons by analogy, from the motion of a tank of water carried on a moving barge (see Galilei 1632/2001, 493). Galileo found this argument so compelling that he initially wanted to title his dialogue *The Discourse on the Tides* (see Shea and Artigas 2003, 125). Ultimately, on the urging of others, Galileo did not to publish the book under that title. Further, he was mistaken about the cause of the tides. Importantly, though, Galileo was a realist about the cause of the tides. Hence, as with Kepler, Galileo's realist convictions extended to his false beliefs as well as those we continue to regard as true.

Galileo played another important role in the Copernican Revolution, one that is especially relevant to the realism/anti-realism debate. In 1616, when Galileo was first urged by the Catholic Church to not teach or defend the Copernican theory, he was told that he could entertain the theory, treating it hypothetically. That is, it was suggested that he could discuss the theory, provided he did so as an instrumentalist (see Special Commission's Report 1632/2008, 273).

In 1632, when Galileo published *Dialogue Concerning the Two Chief World Systems*, he was less cautious than he should have been, and mistakenly thought that he could be more forthright about his realist convictions in astronomy. This was a serious miscalculation. In the *Dialogue*, Galileo has the character Simplicio express the view that God could have created the world in any number of different ways to yield the phenomena that were the basis of astronomers' theorizing about the cosmos (see Galilei 1632/2001, 538). God, after all, is all-powerful, and it is vanity on our part to think we can determine how He constructed the world.

At Galileo's trial by the Inquisition, the Church maintained that the available evidence did not warrant belief in the Copernican theory. Indeed, the Inquisition was concerned that some of the claims entailed by the Copernican theory were either heretical, false, or both (see Inquisition 1633/2008, 292). In the dispute between Galileo and the Catholic Church, what was at issue was what God could or could not do.[17] This issue seems out of place in contemporary scientific debates. But even without appealing to the infinite power of God, many scientists have questioned whether we have adequate grounds for accepting as true hypotheses about the unobservable reality lying behind the observable.

Even though Galileo faced sanctions from the Catholic Church, by the 1630s the Copernican theory, in the modified form developed by Kepler and Galileo, was on its way to becoming the dominant theory in astronomy. The revolution was more or less complete, even though the Church had tried to halt it, and holdouts remained. In fact, the limited reach of the Church in halting the spread of the Copernican theory is evident by the fact that outside of Italy, very few copies of Copernicus' book were censored according to the Inquisition's instructions (see Gingerich 2004, 145–146). As Gingerich notes, not even in France and Spain did owners of the book feel compelled to censor their copies as instructed.

I will return to the examples discussed in this chapter throughout the remainder of the book, as many of the key issues that divide realists and anti-realists in the contemporary debate can be vividly illustrated by this revolutionary change of theory in the history of astronomy.

---

[17] Also at issue was the relationship between science and religion. From the Church's point of view, Galileo was not qualified to adjudicate this dispute. He had no training in theology.

2 | The Underdetermination of Theory Choice by Evidence

I want to begin by clarifying the sorts of logical considerations that threaten scientific realism. The most common logical consideration that is raised against realism is the underdetermination of theory choice by evidence. Roughly, the concern is that there may be other theories that can account for the data as well as the theory scientists currently accept. If this is the case, it raises doubts about whether the accepted theory is true. In this chapter, I want to briefly discuss the relevance of underdetermination to the contemporary debate between realists and anti-realists. The most effective way into this topic is through a brief survey of the history of the issue. The literature on this topic is vast, and much of it is tangential to the concerns I have. Consequently, I will be quite selective in my analysis. I distinguish between various forms of underdetermination, and I argue that the form most relevant to the contemporary debate between realists and anti-realists concerns the possibility of confirming a theory by testing it against a competing theory. I argue that when scientists conduct tests involving competing hypotheses, they are not warranted in inferring that the superior hypothesis is true or approximately true. This insight plays an important role in other anti-realist arguments discussed later.

I begin by examining Pierre Duhem's remarks related to underdetermination. Then I examine W. V. Quine's remarks on underdetermination. Next, I briefly examine the threat of radical underdetermination, a form of underdetermination that owes its inspiration to Quine. Then I comment on the notion of transient underdetermination. Finally, I examine the relevance of underdetermination to the anti-realists' arguments in the contemporary realism/anti-realism debate. I do not think that the underdetermination of theory choice by evidence is the key issue threatening scientific realism.

Duhem and Underdetermination

The underdetermination thesis is often called the *Quine-Duhem Thesis* (see, for example, Fodor and Lepore 1992, 37). This is unfortunate for two reasons. First, Duhem's concerns related to underdetermination are quite different from Quine's concerns. I am not the first to note this (see Fodor and Lepore 1992; Gillies 1993, 98, chapter 2). Second, there are actually a number of different concerns that fall under the label "underdetermination," so it is misleading to speak of *the* underdetermination thesis (see Laudan 1990). Let us begin with a consideration of Duhem's views as they relate to the underdetermination of theory choice by evidence. There are two issues he discusses that are relevant to the topic.

First, Duhem discusses a form of underdetermination that is relevant to falsifying a hypothesis. Duhem defends a form of holism. I will refer to it as "experimental holism," to distinguish it from meaning holism. Duhem argues that when a physicist tests a hypothesis, in the process of conducting the test, he inevitably makes a variety of assumptions by virtue of the accepted theories implicated in the test (see Duhem 1906/1954, 183). According to Duhem, the reason for this is that testing involves deriving a prediction from a hypothesis. But, as Duhem notes, "the predication of the phenomenon, whose nonproduction is to cut off debate, does not derive from the proposition challenged *if taken by itself*" (185; emphasis added). It is only in conjunction with other propositions, drawn from the accepted theories implicated in the test, that the prediction can be derived.[1] As a consequence of this experimental holism, Duhem claims, "if the predicted phenomenon is not produced, not only is the proposition questioned at fault, but so is the whole theoretical scaffolding used by the physicist" (Duhem 1906/1954, 185). The challenge the physicist faces is to determine the source of the error or failure. It could be the hypothesis that is being tested. Alternatively, it could be some other assumption implicated in the test. As Duhem explains,

---

[1] This dependence that Duhem draws attention to, the fact that one can only draw an inference with the aid of additional assumptions, is a central part of Helen Longino's argument for the claim that science is value-laden. Longino believes that values often *inadvertently* affect science through the background assumptions that play a mediating role in inferences (see Longino 1990).

when the experiment is in disagreement with [a physicist's] predictions, what he learns is that at least one of the hypotheses constituting this group is unacceptable and ought to be modified; but the experiment does not designate which one should be changed. (Duhem 1906/1954, 187)

Incidentally, Duhem thought that this problem was a problem for *physicists*. He explicitly claims that this problem is not encountered in "physiology or certain branches of chemistry" (Duhem 1906/1954, 180). Duhem argues that, unlike physics, in these latter fields "the experimenter reasons directly on the facts by a method which is only common sense brought to greater attentiveness but where mathematical theory has not yet introduced its symbolic representations" (180). So it is the intensive use of mathematics and the remoteness from common sense that make physics vulnerable to this problem.

We need not determine (i) whether Duhem is correct about the scope of the problem or (ii) whether these other fields have since reached the level of maturity that characterized physics in Duhem's time. The important issue for our purposes is to note the nature of the problem that concerns him. When one tests a hypothesis and one's prediction is not vindicated, it can be challenging to determine whether the hypothesis is false, or whether some other proposition implicated in the test situation is false. It is the holistic nature of the testing situation that gives rise to this problem.[2]

Duhem suggests that because logic does not dictate how physicists should respond to a recalcitrant observation, they must rely on good sense to guide them. Good sense involves, among other things, impartiality and control over one's "passions and interests" (see Duhem 1906/1954, 218). In exercising good sense, the scientist must guard

---

[2] Karl Popper was aware of the first underdetermination problem that Duhem identified. In *The Logic of Scientific Discovery*, Popper discusses the situation where a scientist encounters an experience that conflicts with a prediction derived from a theory. According to Popper, in such a situation "we falsify *the whole system* (the theory as well as the initial conditions) which was required for the deduction of the . . . falsified statement. Thus it cannot be asserted of any statement of the system that it is, or is not, specifically upset by the falsification" (1935/2002, 56). Popper seems to think that Duhem believes that any hypothesis can be saved come what may. This is apparent from Popper's critical discussion of conventionalism (see Popper 1935/2002, chapter 4). Conventionalists, as Popper understood them, allow the introduction of ad hoc adjustments to a theory in order to account for a recalcitrant experience, a practice that Popper adamantly rejects. I examine Popper's attack on conventionalism in detail in Wray (2015b).

against "vanity which makes a physicist too indulgent towards his own system and too severe towards the system of another" (218). But, even guided by good sense, rational scientists can disagree about how to respond to a recalcitrant experience (see Duhem 1906/1954, 217). Some may respond by rejecting the hypothesis, others by altering the background assumptions. Both types of response can be consistent with good sense. Still, Duhem never took this to be an insurmountable problem for physicists. He explains that

this state of indecision does not last forever. The day arrives when good sense comes out so clearly in favor of one of the two sides that the other side gives up the struggle even though pure logic would not forbid its continuation. (Duhem 1906/1954, 218)

Thus the problem posed by this type of underdetermination is a temporary state in science. Duhem does not believe it provides grounds for a thoroughgoing skepticism.[3]

Duhem raises another concern. This second concern involves a form of underdetermination that is relevant to the confirmation or verification of a hypothesis. Duhem argues that scientists cannot prove a hypothesis is true in the same way one can prove a proposition is true in geometry, by a *reductio ad absurdum*, pitting one hypothesis against a competing hypothesis. Duhem asks, rhetorically, "do two hypotheses in physics ever constitute . . . a strict dilemma?" (Duhem 1906/1954, 190). If the answer is yes, then by proving that one hypothesis entails a false prediction, one would in turn prove the other hypothesis is true. But Duhem believes the answer to his question is no.

Duhem illustrates this with an example from the history of science involving the competition between Newton's theory and Fresnel's theory of the nature of light. Newton believed that "light consisted of projectiles hurled with extreme speed" (Duhem 1906/1954, 189). Fresnel, on the other hand, claimed that "light consisted of vibrations whose waves are propagated within an ether" (Duhem 1906/1954, 189). An experiment that proves that one of these hypotheses is false does not in turn prove that the other hypothesis is true. After all, these

---

[3] Thomas Kuhn's view is similar to Duhem's. Kuhn believes that when a field is in crisis, that is, when no theory holds the allegiance of the whole research community, subjective factors play an important role in ensuring that competing hypotheses are developed. In time, it becomes clear to all or most scientists which is the superior hypothesis (see Kuhn 1977; see also Wray 2011, 160–164).

two hypotheses about the nature of light do not exhaust the possibilities. In fact, as Duhem notes, Maxwell provided an alternative account of the nature of light, "[attributing] light to a periodical electrical disturbance that is propagated within a dielectric medium" (1906/ 1954, 190). And Maxwell's theory was consistent with the evidence elicited from François Arago's attempt to construct a crucial experiment between Newton's and Fresnel's theories. Arago's mistake was to think he had constructed a crucial experiment that would unequivocally determine which of the two competing theories was true, Newton's or Fresnel's.[4]

Duhem argues that "unlike the reduction to absurdity employed by geometers, experimental contradiction does not have the power to transform a physical hypothesis into an indisputable truth" (1906/ 1954, 190). In fact, he claims that "the physicist is never sure he has exhausted all the imaginable assumptions" (190). That is why there are no crucial tests in physics (188). And, consequently, that is why there are no *proofs* in physics as there are in geometry.

This is a different sort of underdetermination problem than the first concern related to experimental holism.[5] The problem here is that scientists are never testing a set of hypotheses that would enable them to *prove* which theory is true, because they are never considering an exhaustive set of hypotheses. What physicists do is construct tests on the basis of the theories they have developed. It is presumptuous, though, to assume that the true hypothesis is among the set of hypotheses being considered in any given case.

Let me underscore the key difference between Duhem's two concerns. Duhem's first concern pertains to the possibility of *falsifying* a hypothesis, whereas his second concern pertains to the possibility of *confirming* a hypothesis. According to Duhem, both falsifying a hypothesis and confirming a hypothesis are more complex processes than many think. Duhem, though, does not take either of these

---

[4] Newton's theory "declares that light travels more quickly in water than in air," and Fresnel's theory "declares that light travels more quickly in air than in water" (see Duhem 1906/1954, 189–190). Arago thought he could construct a crucial experiment by attempting to answer the question: Does light move more quickly in water than in air? But, as Duhem notes, other theories may also be compatible with the results of Arago's experiment, as in fact Maxwell's theory is.

[5] Donald Gillies suggests that this second consideration is a consequence of the first (see Gillies 1993, 101). I disagree, and Duhem regards them as distinct concerns.

challenges to be insurmountable. But they do suggest that physicists should be modest in the conclusions they draw from their tests, whether they are testing a single hypothesis on the basis of a prediction drawn from it or pitting two competing theories against each other. In order to distinguish the two concerns, I will refer to the first concern as "underdetermination with respect to *falsification*" and the second concern as "underdetermination with respect to *confirmation*."[6]

Quine and Underdetermination

Duhem was concerned with the *practice* of science. But with Quine's "Two Dogmas of Empiricism," discussions of underdetermination took a turn away from scientific practice. This is not surprising, given that Quine was a logician, not a practicing scientist, nor a historian of science.

Quine's first mention of the topic of underdetermination in "Two Dogmas" is quite innocuous. He claims that "our statements about the external world face the tribunal of sense experience not individually but only as a corporate body" (Quine 1951, 38). This sounds much like Duhem's experimental holism, though Quine does not restrict the scope of his claim to physics, as Duhem does. In the original article, published in *The Philosophical Review* in 1951, there was no reference to Duhem. Two years later, however, when "Two Dogmas" was reprinted in *From a Logical Point of View*, Quine added a citation to Duhem (see Quine 1951/1953, 41 and Note 17). According to Richard Creath, "Quine ... added the citation to Duhem ... at Hempel's suggestion" (see Creath 2007, 337).[7] So it appears that Duhem was not the source of Quine's thinking about underdetermination. And in fact, as Quine develops his point, it is clear that his notion of under-determination is significantly different from Duhem's.

---

[6] In their discussion of Quine's holism, Fodor and Lepore draw a distinction between Quine's semantic holism and Q-D holism, that is, Quine-Duhem holism. The latter they refer to as "confirmation holism." They characterize confirmation holism in the following way: "every statement in a theory (partially) determines the level of confirmation of every other statement in the theory" (1992, 41). This is not Duhem's view. Duhem's remarks on confirmation are intended to show that you cannot prove a theory is true in the way a geometer can prove a claim is true by constructing a *reductio ad absurdum*.

[7] Gillies mistakenly claims that Quine cites Duhem in the 1951 version of "Two Dogmas" (see Gillies 1993, 108).

According to Quine, "science ... is so underdetermined by ... experience ... that there is much latitude of choice as to what statements to reevaluate in light of a single contrary experience" (Quine 1951, 39–40). Quine compares "total science" to "a field of force whose boundary conditions are experience" (39). The more theoretical claims are likened to the interior of the force field. Quine insists that "no particular experiences are linked with any particular statements in the interior of the field, except indirectly through considerations of equilibrium affecting the field as a whole" (40). Thus, Quine argues that "it is misleading to speak of the empirical content of an individual statement" (40). In speaking this way, Quine shifts the focus from experimental holism to meaning holism.[8] Meaning holism, though, was no part of Duhem's concern.

With this subtle but significant departure from Duhem's view, Quine proceeds to move further in a new direction. He notes that, when faced with a recalcitrant experience, an experience that does not match our expectations, given our beliefs, "any statement can be held true come what may, if we make drastic enough adjustments elsewhere in the system" (1951, 40). Quine is not concerned at all with experimental holism and its effects on scientific practice. Nor is he concerned with the logic of so-called crucial experiments that pit two hypotheses against each other. Rather, Quine's concern is with the *logical* possibilities of dealing with recalcitrant experience, given his meaning holism (see Laudan 1990, 267). That is the focus of his underdetermination thesis. In fact, Quine's concerns have no special connection to science. His account of belief revision in the light of recalcitrant experiences was presented as a characterization of the layperson's predicament, as well as the scientist's.

Radical Underdetermination and Empirically Equivalent Theories

In a comprehensive analysis of "the doctrine of underdetermination," Larry Laudan argues that Quine presents two distinct underdetermination theses. Laudan refers to the one as the *nonuniqueness thesis* and the other as the *egalitarian thesis*. The nonuniqueness thesis states that

---

[8] Fodor and Lepore use the terms "semantic holism" and "meaning holism" interchangeably (1992, chapter 2).

"for any theory, T, and any given body of evidence supporting T, there is at least one rival (i.e. contrary) to T that is as well supported as T" (Laudan 1990, 271; emphasis in original). The egalitarian thesis is stronger. It states that *"every theory is as well supported by the evidence as any of its rivals"* (271). Laudan argues that even the weaker thesis, the nonuniqueness thesis, has not been shown to be well supported. And the stronger thesis, the egalitarian thesis, he claims, is implausible (see Laudan 1990, 275).

The nonuniqueness thesis has dominated discussions of the underdetermination of theory choice by evidence in the last five decades. It is discussed under a variety of different names, including strong underdetermination, radical underdetermination, and global underdetermination.

Lawrence Sklar, for example, characterizes *radical underdetermination* in the following way:

no rational grounds for choosing between two alternative incompatible theories can be found in inductive inference from the data by all reasonable canons of confirmation, rules for inferring to the best explanation, principles of a priori plausibility and so forth. (Sklar 1975, 379; emphasis added)[9]

Similar characterizations of the underdetermination thesis can be found in numerous sources. William Newton-Smith, for example, characterizes the thesis of *strong* underdetermination in the following way: "there could be rival theories that *no data* could decide between; ... all theories are underdetermined by *all actual and possible* observational evidence" (see Newton-Smith 2000, 532; emphasis added). And Justin Biddle characterizes *global* underdetermination in the following way: "*all* theories (or hypotheses, models, etc.) are underdetermined by logic and *all possible* evidence; this leaves a gap between logic and evidence, on the one hand, and theory choice, on the other, which is inevitably filled by contextual factors" (Biddle 2013, 125; emphasis in original). These are merely alternative expressions of the same basic view.

Much of the debate surrounding the nonuniqueness thesis has focused on whether or not there are empirically equivalent rival theories for *every* theory. Frequently, the discussants appeal to contrived

---

[9] Sklar attributes this radical thesis to Descartes, Poincaré, and Quine. Sklar counts Poincaré as a proponent of radical underdetermination by virtue of his views on non-Euclidean geometry (see Sklar 1975, 379).

examples that are far removed from the concerns that motivated Duhem, a practicing scientist. A frequently discussed example is whether Newton's theory, that is, his three laws of motion and the law of universal gravitational attraction, is empirically equivalent to Newton's theory plus "the hypothesis that the center of mass of the universe has constant absolute velocity v" (see, for example, Laudan and Leplin 1991, 457; Earman 1993, 31).

Some have argued that there are ways to generate empirically equivalent theories for any particular theory, thus raising the threat of radical underdetermination for any scientific theory. In fact, Andre Kukla (1996b) argues that there are at least four algorithms for generating empirically equivalent rivals to any theory (see pages 145; 151; 156–157; and 157–158 for a description of the algorithms). Kukla argues that the Underdetermination Argument requires such an algorithm because "it needs to be established that there are empirically equivalent rivals to *any* theory" (1996b, 138).

I am inclined to think that the discussion of empirically equivalent theories was a wrong turn in the debate about underdetermination, at least insofar as it relates to the realism/anti-realism debate. Kyle Stanford provides an insightful assessment of this debate about empirically equivalent theories. He argues that in this debate, the underdetermination problem inadvertently gets pushed aside. Instead, these debates end up focusing on some other philosophical problem that has no "special significance for theoretical science" (Stanford 2006, 13).[10] For example, the debates about the existence of empirically equivalent theories often seem to focus on a skeptical threat not unlike the threat associated with the Cartesian Evil Demon (see Stanford 2006, 12–15).[11] This sort of skepticism, though, is not a central issue in the contemporary realism/anti-realism debate. Contemporary anti-realists are not concerned about such a wide-ranging and corrosive form of skepticism. Rather, they are narrowly concerned with assessing the grounds for skepticism about theoretical knowledge. Thus, much of the literature on underdetermination is irrelevant to the contemporary debate.

---

[10] See Kyle Stanford (2006, 11–13) for a brief analysis of the debate about empirically equivalent theories.

[11] Kukla admits that one of his algorithms "is a minor variant of the Cartesian story about the evil genius" (see Kukla 1996b, 158).

Transient Underdetermination

In the 1970s, Sklar introduced the notion of transient underdetermination, contrasting it with radical underdetermination. As we saw earlier, radical underdetermination is alleged to be a persistent threat, and is alleged to affect all areas of science. Transient underdetermination is quite a different matter. Sklar characterizes transient underdetermination in the following way: "there can be incompatible alternatives between which no rational choice can be made on the basis of a priori plausibilities, strength, simplicity, inductive confirmation, and so forth, *relative to present empirical evidence*" (Sklar 1975, 380–381; emphasis in original). The key here is that it is a temporary threat. Given the evidence currently available, neither of two competing theories seems unequivocally superior. The assumption is that if we gather more evidence, then the underdetermination might pass; it was only transient. Though Sklar drew attention to transient underdetermination in the mid-1970s, most philosophers of science writing about the issue of underdetermination until quite recently were preoccupied with radical underdetermination, what Laudan calls the nonuniqueness thesis.

Recently, though, philosophers of science have considered the specific challenges raised by transient underdetermination. Both Stanford and Biddle, for example, have recently discussed transient underdetermination (see Stanford 2006, 17; Biddle 2013, 125). Biddle claims that the thesis of transient underdetermination is "undoubtedly true" (125). Scientists are sometimes confronted with a choice between two competing theories that are equally well supported by the available data.

In an effort to be clear about the nature of this form of underdetermination, it is worth examining an example of transient underdetermination drawn from the history of astronomy. Recall from Chapter 1, in the mid-1500s, European astronomers were faced with a choice between two well-developed competing theories, Copernicus' heliocentric theory and the late-Renaissance version of Ptolemy's geocentric theory. These theories differed not only in the cosmology they assumed, but also in their success. But astronomers could *reasonably* accept either theory, given the evidence then available. Each of the competing theories was plausible, and the two theories were comparable with respect to the empirical support they had. By the late 1580s, astronomers had a third contender, Tycho Brahe's theory.

The underdetermination of theory choice in this case, though transient, was quite persistent. Arguably, it was not until the second decade of the 1600s that the Ptolemaic theory was no longer regarded as a serious contender. As noted earlier, Galileo's telescopic observations, especially the phases of Venus, contributed significantly to undermining astronomers' allegiance to the Ptolemaic theory. But it took longer for the battle between the Tychonic theory and the Copernican theory to be resolved, as Galileo's observations were compatible with both theories.

The various theories that astronomers considered during the late sixteenth and early seventeenth centuries are not empirically equivalent. They entail radically different predictions about the world, contrary to what we would expect from empirically equivalent theories. Those who are concerned with transient underdetermination are not concerned that two competing theories entail all the same observable consequences. The significance of transient underdetermination is that it captures a type of situation that scientists sometimes actually experience. Sometimes scientists are not able to determine unequivocally which of two competing theories is superior, given the data they have access to. This is a type of underdetermination that we should take seriously.

Duhem's first concern, the concern related to underdetermination with respect to falsification, is a form of transient underdetermination. It is a temporary problem in a scientific field, but one that is resolved in a principled way when more data are gathered.

But there is nothing about transient underdetermination that speaks either in favor of anti-realism or against it. This form of underdetermination is thus tangential to the contemporary debate between realists and anti-realists. Contemporary anti-realists are not merely claiming that sometimes scientists are unable to determine which of two competing theories is superior. The sorts of concerns that motivate contemporary anti-realists are more pressing and persistent.

Underdetermination and the Contemporary Realism/Anti-Realism Debate

In this section, I want to briefly explain the relationship between Duhem's concern about underdetermination with respect to confirmation and the arguments that have been advanced by contemporary anti-realists.

The type of underdetermination most relevant to the contemporary debates is the second form that Duhem discusses, what I call "underdetermination with respect to confirmation." Recall that Duhem suggests that tests of competing scientific hypotheses are not like *reductio ad absurdum* proofs in geometry, where proving that one of the hypotheses is false amounts to proving that the other hypothesis is true. When scientists conduct tests to determine which of the competing hypotheses is superior, they are not considering an exhaustive set of hypotheses. As a result, they are not in a position to infer that the hypothesis that survives such a test is likely true or approximately true. This insight figures in two of the most influential arguments in the contemporary realism/anti-realism debate.

First, as we will see in Chapter 5, Stanford's Argument from Unconceived Alternatives, the New Induction over the History of Science, is based on this insight. Stanford argues that reflection on the history of science shows that scientists have developed theories that can account for the data that supported theories that were accepted earlier, but these alternative theories were unconceived at the time the earlier theory was the accepted theory (see Stanford 2006). This, he argues, is a typical occurrence. Hence, scientists should be aware that even today's best theories may be replaced in the future by yet unconceived theories.

Second, as we will see in the next chapter, Bas van Fraassen's Argument from a Bad Lot, the Argument from Underconsideration, is also based on the same sort of underdetermination (see van Fraassen 1989). Scientists, van Fraassen argues, are choosing between sets of theories that only include those developed to date. Usually, there are only ever two or three well-developed alternatives competing for the allegiance of scientists at any given time. Like Duhem, van Fraassen thinks that it is hubris on the part of scientists to think that the true theory is among the set of theories from which they are choosing. Because of this sort of underdetermination, any inference from a test of a theory to its truth or approximate truth is unwarranted. And there is no reason to think that this form of underdetermination only affects physicists. Rather, all scientists seem to be confronted by this same problem when they are choosing between competing theories.

One final remark is in order about the difference between Quine's concern and Duhem's concern regarding underdetermination with respect to confirmation. The underdetermination thesis associated with Quine emphasizes the fact that scientists lack sufficient data to discern between competing theories, actual competitors, and logically possible competitors. Quine contrasts the meager input from the senses with the torrential output of the content of our theories. The content of our theories, he notes, far surpasses what we learn from the senses (see Quine 1969, 83). Thus, it is the paucity of data that creates the problem. Duhem, on the other hand, emphasizes the fact that there is a shortage of theories. And this is what undermines any inference in a test situation to the truth of a theory. As long as scientists are choosing between just a few competitors, an inference to the truth of the superior theory will be unwarranted.

This analysis of the underdetermination of theory choice by evidence draws attention to a key challenge that any form of scientific realism faces. Realists have been preoccupied with showing that scientists can confirm theories, even as they acknowledge scientists' fallibility. But this narrow focus on advancing our understanding of confirmation fails to address a key issue that Duhem draws our attention to. Methods of confirmation can be quite rigorous, but unless scientists have developed a true theory (or approximately true theory), their methods of confirmation cannot deliver what realists are seeking. The methods of confirmation that are widely discussed in the realism/anti-realism debate merely provide a means for evaluating the various theories developed. They do not provide a means for generating new theories. It seems clear that the context of discovery, including the methods by which scientists generate new theories and hypotheses, is relevant to understanding scientists' prospects of getting at the truth. This topic, though, has largely been neglected in the contemporary realism/anti-realism debate.

3 | The Argument from Underconsideration

The anti-realists' Argument from Underconsideration is far less discussed in the literature than the Argument from Underdetermination, despite the fact that anti-realists have appealed to the argument for some time (see, for example, van Fraassen 1989, 142–150). This argument focuses on the fact that when scientists evaluate theories, they only ever consider a subset of the theories that can account for the available data, specifically those theories that have been developed. Indeed, when scientists are evaluating theories, typically they are choosing between two or three competitor theories. As a result, the anti-realist argues, when a scientist judges one theory to be superior to competitor theories, she is hardly warranted in drawing the conclusion that the superior theory is likely true with respect to what it says about unobservable entities and processes. Anti-realists claim that the inference to the likely truth of the superior theory presumes that scientists are especially skilled at developing theories that are true. But the history of science seems to suggest otherwise. Scientists do not have such epistemic privilege.

Realists are not compelled by this argument. Peter Lipton (1993/ 1996), for example, argues that the Argument from Underconsideration fails to provide adequate support for its conclusion. Specifically, Lipton argues that the premises of the Argument from Underconsideration are inconsistent, and one premise is false. In this chapter, I defend the Argument from Underconsideration. I argue that Lipton is mistaken in his assessment of the Argument from Underconsideration. Thus, the argument remains a genuine threat to scientific realism. Indeed, I think it is one of the most compelling arguments in support of anti-realism.

I begin by presenting the Argument from Underconsideration and Lipton's criticisms of it. Then I clarify the nature of the *reliability* that anti-realists attribute to scientists in the Argument from Underconsideration. Anti-realists are *not* radical skeptics. They do not believe that scientists are wholly unreliable. Indeed, they are impressed with science and the accomplishments of scientists, though they disagree with realists about what the real accomplishments of science are. I then argue that the alleged inconsistency that Lipton claims to find in the Argument from Underconsideration vanishes once we understand what the anti-realist means when she claims that scientists are generally reliable. Then I take issue with a particular strategy Lipton employs in his efforts to defend realism against this argument. I argue that collapsing relative evaluations of theories into absolute evaluations of theories, as Lipton recommends, has its costs. Specifically, a test no longer proves that a single theory is true, but rather that the truth is among a family of theories. Moreover, the realist is in no position to determine which theory in that family of theories is the true theory. Thus, I argue that Lipton's strategy is counterproductive. Finally, I briefly examine Richard Boyd's (1983; 1985) influential defense of realism. I argue that it is based on assumptions similar to those that motivate Lipton in his critique of the Argument from Underconsideration, assumptions that I aim to show are flawed.

The Argument from Underconsideration and Lipton's Concerns

Let me begin by explaining the Argument from Underconsideration. I will focus on Lipton's reconstruction of the argument. In fact, I believe that Lipton's reconstruction of this anti-realist argument is a fair reconstruction. As we will see shortly, it is his evaluation of the argument that I object to.

According to Lipton, the Argument from Underconsideration consists of two premises, which he calls the Ranking Premise and the No-Privilege Premise. According to the Ranking Premise, evaluations of theories give "only a comparative warrant" (Lipton 1993/1996, 93). That is, when two theories are compared, one theory is judged to be superior to the competitor, rather than *categorically* true (see van Fraassen 1989, 142–143).[1] According to the No-Privilege Premise, "scientists have no reason to suppose that ... it [is] likely that a true theory will be among [the set of theories from which they are

---

[1] Thomas Kuhn did much to popularize the notion that theory evaluation is comparative (see Kuhn 1962/2012). He argues that scientists do not just compare

choosing]" (Lipton 1993/1996, 93). Indeed, to suppose that the true theory is among the set is to suppose that contemporary scientists are privileged in their access to unobservables (see van Fraassen 1989, 143–144). The conclusion the anti-realist draws is that scientists are not warranted in inferring that the superior theory is likely true (Lipton 1993/1996, 94). Though the superior theory is more likely true than the competitor theories with which it is compared, it does not follow that the superior theory is more likely true than not. As Bas van Fraassen (1989) puts the point, scientists may merely be choosing the best of a bad lot (149). A bad lot is just a set of theories that does not include a true theory in it.

Importantly, Lipton notes that the anti-realists who advance this argument are not presenting a thoroughgoing skeptical argument. Rather, according to Lipton, the anti-realists who endorse this argument grant that "scientists can rank the competing theories they have generated with respect to the likelihood of truth" (Lipton 1993/ 1996, 93). Hereafter, I will refer to this claim as the *Reliability Assumption*. The Reliability Assumption is an implicit premise in the argument. It is worth making it explicit, because this assumption figures importantly in one of Lipton's criticisms of the Argument from Underconsideration.

Let us now consider the two concerns Lipton has with this argument. First, Lipton believes that the Ranking Premise is false. Second, he claims that the Ranking Premise and the No-Privilege Premise are inconsistent. Given these problems, Lipton believes that we should not accept the conclusion, at least not on the basis of *this* argument.

Let us begin with his concern with the first premise. Contrary to what is suggested by the first premise, Lipton believes that any relative evaluation can be collapsed into an absolute evaluation, for every "pair of contraries entails a pair of contradictories" (1993/1996, 98). Contraries are pairs of claims such that if we know one is true, then we know the other is false, but if we know one is false, we do not necessarily know the other is true. For example, the following two claims are contraries: the Earth is at the center of the cosmos, and the Sun is at the center of the cosmos. If I know one of these claims is true,

theories to the world. Rather, theory evaluation involves comparing competing theories with each other and the world.

I know the other is false. After all, the Earth and the Sun cannot both be at the center of the cosmos. But if I know one of the claims is false, I cannot infer that the other is true. After all, it may be the case that neither the Earth nor the Sun is at the center of the cosmos. Contradictories, on the other hand, are pairs of claims such that if I know one is false, then I know the other is true, and if I know one is true, then I know the other is false. The following two claims are contradictories: the Earth is at the center of the cosmos, and the Earth is not at the center of the cosmos. The truth of one of these claims entails the falsehood of the other, and the falsity of one entails the truth of the other. Contradictories, unlike contraries, exhaust the possibilities.

Let us examine how Lipton proposes to collapse relative evaluations into absolute evaluations in science. He refers to theories in the abstract as T1, T2, etc. According to Lipton,

all pairs of contraries entail a pair of contradictories, since one member of such a pair always entails the negation of the other. Suppose ... we wish to rank the contradictories T1 and ~T1. If we find a contrary to T1 (say T2) that is ranked ahead of T1, then ~T1 is ranked ahead of T1, since T2 entails ~T1. Alternatively, if we find a contrary of ~T1 (say T3) that is ahead of ~T1, then T1 is ranked ahead of ~T1, since T3 entails T1. (1993/1996, 98)

The key to generating theories that are contradictories from two theories that are merely contraries is to find the point at which one theory denies some claim of the other. With respect to *that* claim, the theories are contradictories. For example, though strictly speaking Ptolemy's theory of the cosmos and Tycho Brahe's theory of the cosmos are merely contraries, with respect to the claim "Venus orbits the sun" (H1), they are contradictories. Tycho's theory maintains that Venus orbits the sun, whereas Ptolemy's theory maintains that Venus does not orbit the sun (~H1). Hence, whatever other similarities or differences there may be between the two theories, with respect to this particular claim they are contradictories. Hence, a relative evaluation of these two competing theories can be reframed as an absolute evaluation of the two competing contradictories, H1 and ~H1. In a similar manner, Lipton argues, any comparative or relative judgment can be reconfigured or collapsed into an absolute judgment. Hence, Lipton concludes that the ranking premise is false. Scientists can and do make absolute evaluations.

Lipton discusses a particular context in which scientists routinely make absolute evaluations. When testing whether some factor has a causal impact on the effect they are studying, psychologists frequently first aim to show that the null hypothesis is false. The null hypothesis states that the factor in question has no impact. For example, a psychologist might be interested in determining whether some sort of intervention, like an after-school reading program, has an impact on children's performance on standardized tests. The null hypothesis says that the program has no effect. The psychologist might conduct a test to determine if the data support the null hypothesis. In testing the null hypothesis, it seems that psychologists are in fact making an absolute judgment: either the factor in question has no causal influence, or it has some causal influence.

Let us consider Lipton's second concern with the Argument from Underconsideration. He argues that "the two premises of the argument from underconsideration are incompatible" (1993/1996, 100). The alleged incompatibility to which Lipton draws attention does not concern the Ranking Premise and the No-Privilege Premise, but rather the *Reliability Assumption* and the No-Privilege Premise. Lipton argues that if scientists are reliable in their evaluations, as the anti-realist assumes, then, contrary to what the No-Privilege Premise suggests, we must assume that scientists are more apt than not to have the true theory among the set of theories from which they are choosing. Lipton believes that if scientists were generally unable to generate true theories, then they would not be very reliable. And if they were not reliable, then, faced with a choice between two theories, we would have little reason to believe that they are capable of reliably choosing the theory that is superior. But given the reliability of scientists, we have reason to believe that they are generally apt to be choosing from a set of theories that includes the true theory.

To understand the grounds for Lipton's criticism, we need to consider the role played by background theories in evaluations or tests of competing theories. As Lipton notes, scientists do not evaluate competing theories in isolation. Rather, they "rank new theories with the help of background theories" (1993/1996, 100). Background theories thus play an indispensable role in science. Lipton explains that background theories "influence the scientists' understanding of the instruments they use in their tests, the way the data themselves are to be characterized, the prior plausibility of the theory under test, and [the] bearing of data on the theory" (100).

Lipton claims that scientists' comparative judgments are not aptly described as reliable if their background theories are not generally true. After all, if scientists are relying on false background theories when making their evaluations of competing theories, there is little reason to believe that the conclusions they draw are likely true. And, if this were the case, it would not be surprising that scientists are unreliable. Indeed, as far as Lipton is concerned, if this were the case, it would not be surprising when scientists rank an inferior theory over a superior theory. Alternatively, if the background theories scientists accept are generally true, as the Reliability Assumption implies, then they are not likely to be choosing between theories that are all far off the mark. Their dependence on true background theories will ensure that the theories they develop and ultimately choose are likely true. Hence, the No-Privilege Premise must be false.

Lipton argues that the anti-realist is caught in a dilemma that threatens the Argument from Underconsideration. If the anti-realist wants to maintain that scientists are reliable in their evaluations, then she must admit that they are privileged with respect to their evaluations of theories, both background theories and other theories. But then the anti-realist's skeptical conclusion is unwarranted. Alternatively, if the anti-realist wants to maintain that scientists are not privileged, and hence are apt to be choosing from a bad lot, then there is no basis for claiming that they are reliable with respect to their evaluations. Hence, either the No-Privilege Premise is false, or the Reliability Assumption is false. As Lipton puts the point, "that scientists might be completely reliable rankers and yet arbitrarily far from the truth is an illusion" (1993/1996, 101). Hence, Lipton suggests that the sort of intermediate skeptical position that anti-realists aim to defend is untenable. Either we must embrace a more thoroughgoing form of skepticism than the anti-realist wants us to accept, or we must admit that the anti-realist's skeptical worries are unjustified.

Given the two problems outlined above, Lipton believes that the Argument from Underconsideration does not provide adequate support for the conclusion. Scientists are sometimes warranted in inferring that the superior theory of two competing theories is true or likely approximately true.

Anti-Realism and Evaluative Reliability

Lipton is correct to claim that anti-realists, and in particular those anti-realists who endorse the Argument from Underconsideration, assume that scientists are generally reliable in their evaluations. After all, as noted earlier, these anti-realists are not thoroughgoing skeptics. Anti-realists are skeptical, but only in a circumscribed manner. Specifically, they are skeptical about:

(I) the claims our theories make about unobservable entities and processes (see, for example, van Fraassen 1980), and

(II) the claim that we have good reason to believe that the true theory is among the set of theories scientists are choosing from (see, for example, van Fraassen 1989, 142–143; also Stanford 2006).

In this section, I want to clarify exactly what it is that the anti-realist grants when she claims that scientists are generally reliable.

The sort of reliability that the anti-realist assumes scientists have is reliability with respect to their judgments of those features of theories that they can ascertain directly, like *predictive accuracy*. Hence, insofar as scientists are reliable, they are reliable in their judgment that T1 is more accurate than T2. Granted, the fact that one theory is more accurate than another gives us *some* reason to believe that the more accurate theory is more likely true than the less accurate theory. But the sort of reliability that the anti-realist assumes scientists have *does not* entitle one to infer that "T1 is likely true" from the judgment "T1 is more accurate than T2." After all, one might be choosing from a bad lot. Without independent reasons for believing that scientists are choosing from a good lot, that is, that the true theory is among the set of theories scientists are choosing from, scientists cannot infer that the more accurate theory is also likely the true theory.

There are, in fact, specific features that some scientists (and some philosophers of science) often *regard* as reliable indicators of the truth of theories. These are typically referred to as theoretical values. Simplicity and breadth of scope, for example, are often regarded as such indicators. The anti-realist can grant that scientists are also generally reliable with respect to their judgments about which of two theories is simpler. And the anti-realist may even grant that the simpler theory is more likely true than the more complex theory with which it is compared.[2] But what the anti-realist denies is that we have good reason to believe that such a judgment warrants the conclusion that the simpler theory is therefore *likely the true theory*, or even likely approximately true. A theory being more likely true than another theory is quite a different matter than a theory being more likely true than not. Unless the true theory is among the set of theories one is comparing, one cannot reasonably conclude that the theory that is most likely true of the lot is also more likely true than not.

The anti-realist can raise similar concerns with respect to other theoretical virtues, breadth of scope and fruitfulness, for example. Though scientists may be reliable in their judgments about the relative superiority of a theory with respect to any theoretical virtue, such judgments do not warrant the conclusion that the superior theory is likely true. At best, such a judgment could support the conclusion that one theory is more likely true than the other theories with which it is compared. But, unless we either (i) can collapse comparative evaluations into absolute evaluations or (ii) have good reason to believe that the true theory is among the set of theories scientists are choosing from, we cannot reasonably infer that the superior theory is likely true with respect to what it says about unobservables. In the remainder of this chapter, I aim to show that (i) there are significant costs to collapsing comparative evaluations into absolute evaluations, costs that threaten Lipton's criticism of the Argument from Underconsideration and (ii) we are not warranted in believing that scientists are generally choosing between sets of theories that contain the true theory.

I will return to a more sustained analysis of the so-called theoretical values and their role in theory evaluation in Chapter 8.

The Alleged Incompatibility and the No-Privilege Premise

Now that I have clarified what it is that the anti-realist means when she grants that scientists are reliable in their evaluations, we are in a better

---

[2] Judgments of relative simplicity are rather complex. For example, as Kuhn (1977) notes, each of two competing theories may be simpler than the other, but in different respects. And there is the further concern that we really do not know what the relationship is between the simplicity of a successful theory or model and the unobservable structure of the world (in this regard, see Cartwright 1983; van Fraassen 1989, 147–148). I will not pursue these concerns in this chapter.

position to evaluate Lipton's two criticisms of the Argument from Underconsideration. My aim is to show that the Argument from Underconsideration is more compelling than Lipton has led us to believe. In this section, I want to examine Lipton's second criticism, that the No-Privilege Premise and the Reliability Assumption are inconsistent. I believe that this alleged inconsistency dissolves once we have a clear understanding of what the anti-realist means by "reliable" when she claims that scientists are reliable with respect to their evaluation of theories.

As we saw above, the sort of reliability that the anti-realist assumes scientists have is with respect to their judgment that one theory is simpler than another, or one theory is more accurate than another. And as far as the anti-realist is concerned, our background theories are no different than other theories in this regard. The only evidence scientists have for the truth of the background theories is their relative accuracy, their relative simplicity, their relative breadth of scope, their relative consistency, and their relative fruitfulness. Hence, strictly speaking, all that scientists can claim to know about the background theories is that they (i) save the phenomena, (ii) embody the various theoretical virtues to some extent, and (iii) are superior to the theories with which they are compared.

Clearly, having *false* but *empirically adequate* background theories is consistent with the sort of evaluative reliability defended in the previous section. An empirically adequate theory is one that accounts for the phenomena or appearances. Such a theory *could* be false with respect to what it says about the underlying structure of the world. So, contrary to what Lipton would have us believe, the following two claims are not incompatible: (i) scientists are reliable in their judgments of competing theories, and yet (ii) they may be working with background theories that are false. Consequently, contrary to what Lipton suggests, the Reliability Assumption is not incompatible with the No-Privilege Premise. Scientists, though reliable in their judgments, may often be choosing the best of a bad lot. So even when they choose the best theory from the lot, they may choose a false theory.

Thus, given that the No-Privilege Premise and the Reliability Assumption are compatible, it seems clear that the anti-realist's intermediate skepticism does not necessarily collapse into either realism or a thoroughgoing skepticism, as Lipton suggests. Hence, contrary to what Lipton claims, the intermediate skepticism that anti-realists recommend is a viable alternative.

Comparative and Absolute Evaluations

Let us now consider Lipton's first criticism of the Argument from Underconsideration. Lipton argues that relative evaluations of theories can be collapsed into absolute evaluations. More precisely, Lipton claims that by recasting our comparative judgments of contraries as absolute judgments of contradictories, we can infer that the theory we judge to be superior is most likely true. I aim to show that Lipton fails to see the price scientists must pay when they collapse comparative evaluations into absolute evaluations as he recommends.

Before we consider the concern I have with collapsing comparative judgments, let us examine the nature of comparative theory evaluations. Anti-realists believe that evaluations have the following form: T1 is superior to T2 with respect to quality A. There are a number of things to note about this type of evaluation. First, this is not a claim to the effect that T1 is *more likely true* than not true. Indeed, it is not a claim about truth except insofar as one believes that a theory having quality A is a reliable indicator of its being true. Second, because we can and do evaluate competing theories with respect to numerous different qualities, including accuracy, simplicity, breadth of scope, internal and external consistency, and fruitfulness, comparative evaluations are apt to be quite complex (see Kuhn 1977, 321–322). Our comparative evaluation may lead us to the conclusion that T1 is superior to T2 with respect to qualities A, B, and C, but inferior with respect to qualities D and E. Given the nature of comparative evaluation, it is not clear that relative evaluations can be so readily collapsed into absolute evaluations. At any rate, the relative evaluation that "T1 is superior to T2 with respect to qualities A, B, and C, but inferior with respect to qualities D and E" cannot be easily reformulated into an absolute evaluation of the form "T1 is superior to ~T1." Moreover, the comparative evaluation does not support the absolute judgment that "T1 is likely true."

Let us suppose that one theory is unequivocally superior to another. That is, let us suppose that one theory beats the competitor theories on all measures. Still, it does not follow that the superior theory is true. Again, unless we have independent reasons for thinking that scientists

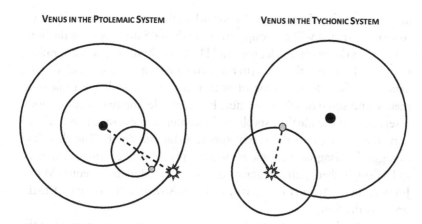

VENUS IN THE PTOLEMAIC SYSTEM   VENUS IN THE TYCHONIC SYSTEM

Figure 2 Venus in the Ptolemaic system and Venus in the Tychonic system

are choosing from a set of theories that includes the true theory, we can only infer that the superior theory is more likely true than the competitors with which it is compared. Such comparative evaluations do not permit us to assign a probability to the truth of the superior theory. We can only make a judgment to the effect that "X is more probable than Y." But "X is more probable than Y" would be true even if the probability of X is only 10 percent, provided the probability of Y is less than 10 percent.

Now, suppose that we grant Lipton his claim that we are able to collapse comparative judgments into absolute judgments. Consider, for example, an early modern astronomer comparing the Tychonic theory and the Ptolemaic theory. In an effort to render these contrary theories into contradictories, the astronomer construes the Tychonic theory as the claim that Venus orbits the sun (H1), and the Ptolemaic theory as the claim that Venus does not orbit the sun (~H1).

These illustrations provide representations of the orbit of Venus in each of the two theories (see Figure 2). In the Ptolemaic theory, Venus is on an epicycle, and the center of Venus's epicycle always lies on the line running from the center of the Earth to the center of the Sun. Recall that this constraint was introduced to account for the fact that Venus is never seen more than 45 degrees from the Sun. That is, it was an ad hoc adjustment to ensure that Ptolemy's model of Venus could save the phenomena. In the Tychonic model, the constrained orbit of Venus is accounted for by the fact that Venus orbits the Sun. Consequently, seen from the Earth, it will never appear more than 45 degrees from the Sun.

For the sake of clarity, let me call H1 the Tychonic *hypothesis*, rather than the Tychonic theory. After all, the Tychonic theory includes an array of other claims, some of which are shared with the Ptolemaic theory and some of which are not. For example, the two theories both assert that (i) the Earth is stable at the center of the cosmos and (ii) the Sun, the Moon, and the starry sphere orbit the Earth. The theories, though, do disagree on some important points. For example, whereas in Ptolemy's theory the center of the orbits of Mercury, Venus, Mars, Jupiter, and Saturn is the Earth, in Brahe's theory the center of their orbits is the Sun.

Let us assume that, on the basis of some test, an astronomer is able to make the absolute judgment that the Tychonic hypothesis is more likely true than the Ptolemaic hypothesis. What Lipton must realize is that he has gained very little by collapsing the comparative evaluation of these theories into an absolute evaluation. Even granting that the Tychonic *hypothesis* is more likely true than the Ptolemaic hypothesis, we cannot now infer that the Tychonic *theory in all its details* is likely true. After all, we really only compared the theories with respect to the claim that Venus orbits the sun. But the claim that Venus orbits the sun is compatible with a number of other theories (some of which have yet to be conceived). Indeed, it is compatible with the Copernican theory, which was already conceived at the time, as well as the so-called Egyptian theory, according to which Venus and Mercury orbit the Sun, but the Sun and the remaining planets orbit the Earth.

Hence, Lipton's strategy of collapsing relative judgments to absolute judgments is achieved at great cost. He is now really only able to draw an inference about the truth of the superior contradictory, Venus orbits the Sun (H1), or Venus does not orbit the Sun (~H1). He is not able to draw a warranted inference about the truth of the competing *theories* in all their rich detail. In fact, he must realize that H1 no longer represents a single theory, but rather a family of theories, most of whose members have yet to be conceived. So he seems to be back where he started. He knows that one theory is superior to another, but he is in no position to know whether that theory is likely true. Thus the strategy of collapsing contraries to contradictories in order to render comparative judgments of theories into absolute judgments of theories fail.

Recall Lipton's appeal to the testing of the null hypothesis discussed earlier. Clearly, when a psychologist shows that the null hypothesis is false, she has proved something of quite limited value. The psychologist cannot specify what the effect of the factor in question is. All she knows is that it plays some causal role. This is rather limited knowledge, and is certainly compatible with a range of hypotheses about the causal powers of the factor in question. Thus, it is somewhat misleading to construe the testing of the null hypothesis as a case where scientists are making absolute judgments or evaluations of competing theories.

Boyd's Defense of Realism

There are interesting parallels between Boyd's defense of realism and Lipton's criticism of the Argument from Underconsideration. Both Boyd and Lipton believe that the anti-realist cannot reconcile (i) the reliability of scientists and (ii) the role of background theories in theory evaluation and testing with (iii) skepticism about the likely truth of our current best theories. Hence, before concluding this chapter, it is worth briefly examining what implications the arguments I developed above have for Boyd's defense of realism (see Boyd 1983; 1985).

Boyd's claim is that only realism can offer an adequate explanation for the instrumental reliability of "the methodological practices of science" (1985, 13). As Boyd explains, "both scientific realists and (almost all) empiricists agree that [the methodological] practices [of contemporary science] are instrumentally reliable" (13). Here we have the analogue of the Reliability Assumption, but Boyd attributes the reliability to the methods used in science rather than to scientists. Also, Boyd notes that our current methodological practices depend for their reliability on the various background theories scientists accept. Unless scientists' background theories are approximately true, Boyd claims, it is unlikely that their methods would be reliable.

Given the dependence of scientists' reasoning on background theories and the reliability of their methods, Boyd believes that it is highly unlikely that scientists will routinely accept false theories. As Boyd explains, "the fact that a proposed theory is inductively supported at the theoretical level on the basis of already confirmed theories constitutes (some) evidence in favor of its approximate truth" (28). Boyd thus believes that the anti-realist's skepticism about the truth of our best theories is incompatible with the reliability of the methods scientists employ (31).

Boyd, though, is making the same mistaken assumption about the nature of instrumental reliability that Lipton makes. Scientists are reliable in determining which of two theories is more accurate, and which of two theories is simpler. That much the anti-realist grants. And scientists' *methods* are reliable in aiding them in determining which theory is simpler or more accurate. But this reliability does not provide adequate evidence for the claim that our current best theories are likely true with respect to what they say about unobservables, nor even the claim that our current best *background* theories are likely true with respect to what they say about *unobservables*. Indeed, one can be confident that our best background theories save the phenomena. After all, that is the reason they were accepted in the first place. That is why we are working with these background theories rather than some other background theories. But a theory can embody the marks of a good theory, and embody them more than the competitor theories it is compared with, and still, in time, turn out to be false. As anti-realists have noted, the history of science is filled with theories that embodied various theoretical virtues and yet have since been rejected.[3]

Hence, contrary to what realists would have us believe, our success in developing predictively accurate theories is compatible with our acceptance of theories that are false with respect to what they say about unobservables. This is only paradoxical to those who are unprepared to accept that false theories can be predictively accurate. The history of science, though, shows that it is not impossible to derive true predictions from false theories. As Larry Laudan notes, for example, scientists were able to accurately predict a wide variety of phenomena using Newton's optical theory, even though the theory "was committed to a basic ontology of light which (so we now believe) is desperately wide of the mark" (1984b, 91). Indeed, as we will see later, false theories have even enabled scientists to generate vindicated predictions of *novel* phenomena (see Carrier 1991; Lyons 2002; Vickers 2013).

In summary, I have shown that Lipton is mistaken in his assessment of the Argument from Underconsideration. First, I have argued that

---

[3] Even Laudan grants that scientists are generally reliable with respect to methods (see Laudan 1984a, 25). But he argues that methodological reliability is compatible with skepticism about the truth of our theories.

collapsing relative evaluations into absolute evaluations has its costs. Though we can learn which of two contradictories is true, we must realize that standing behind each contradictory is not a single theory, but rather a *family of theories*, many of which have yet to be conceived. And it is no easy matter determining which member of the family is the true theory. Hence, we are still unable to justifiably conclude that the superior theory is the true theory.

Second, I have argued that, contrary to what Lipton claims, the No-Privilege Premise and the Reliability Assumption are not incompatible. Lipton was led to think otherwise because he assumes that if the Reliability Assumption is true, then scientists must be reliable with respect to their background theories. And Lipton believes that if scientists are reliable with respect to background theories, then they are probably choosing from a good lot, a set of theories that contains a true theory. The anti-realist grants that scientists are reliable in their judgments of *relative* accuracy, breadth of scope, consistency, etc. But I have argued that there is little evidence that their reliability extends to choosing the theory that best describes reality at the level of the unobservables. Further, for Lipton to assume that scientists are reliable in this way, as he does in his argument against the anti-realist, is to beg the question. History has shown, over and over again, that the theory that is judged at one time to be the most accurate can, in time, prove to be false. We will examine the role that evidence from the history of science has in the realism/anti-realism debate in detail in Chapters 5 and 6. As things stand now, the Argument from Underconsideration remains strong.

4 | *Epistemic Privilege*
Another Realist Dogma

In the previous chapter, we saw that some realists claim that scientists in mature fields are prone to develop theories that are true or approximately true. This amounts to ascribing some sort epistemic privilege to scientists.[1] As we saw, Peter Lipton appeals to this privilege that scientists are alleged to have in order to rebut the anti-realist's Argument from Underconsideration. Other realists have made similar appeals to some sort of epistemic privilege in an effort to explain the success of our current best theories. These realists insist that we have good reason to believe that scientists are even successful in their pursuit of theoretical knowledge, that is, knowledge of unobservable entities and processes. Such defenses of realism have been advanced by Richard Boyd (1983; 1985), Lipton (2004; 1993), Stathis Psillos (1999), and, most recently, J. D. Trout (2016).

In this chapter, I examine the case for epistemic privilege as it is advanced by realists. I argue that the arguments for epistemic privilege are generally quite weak. And whatever epistemic privilege scientists may have, it is not so robust as to ensure that we have good reasons to think that they are inclined to develop true theories or even approximately true theories. If scientists are in fact successful in this endeavor, it is not because they are epistemically privileged.

---

[1] It is worth distinguishing the sense of epistemic privilege assumed here from the sense of epistemic privilege that figures in debates in feminist epistemology. As Kristina Rolin (2006) explains, feminist standpoint theorists defend a thesis of epistemic privilege according to which "underprivileged social positions are likely to generate perspectives that are 'less partial and less distorted' than perspectives generated by other social positions" (125). Epistemic privilege is thus alleged to be a function of a lack of social privilege. This is not the sense of epistemic privilege at stake in this debate.

The Case for Epistemic Privilege

Let us examine the evidence and arguments offered in support of scientists' alleged epistemic privilege. Two types of strategies have been employed, a direct strategy and an indirect strategy. Let us consider the indirect strategy first.

As we saw in the previous chapter, Lipton (1993) discusses epistemic privilege in his critical appraisal of the anti-realists' Argument from Underconsideration. But rather than arguing for epistemic privilege directly, Lipton argues that the anti-realists' assumption that scientists lack such privilege conflicts with the reliability that both realists and anti-realists assume scientists have (2004, 158). As we saw in the previous chapter, Lipton assumes that one will accept that scientists have epistemic privilege when one sees there is a conflict between the No-Privilege Premise and the Reliability Assumption, that is, the claim that scientists are generally reliable in their comparative assessments of competing theories. Denying the latter claim would lead to a thorough-going skepticism, something even the anti-realist does not want. I criticized this argument in detail in the previous chapter, so we need not review the weaknesses of this argument for privilege here.

Psillos (1999) also defends the claim that scientists are epistemically privileged. Like Lipton, Psillos develops an indirect argument. Instead of presenting independent evidence to support the claim that scientists are epistemically privileged, Psillos argues that both realists and anti-realists need to *assume* that scientists are privileged. The disagreement between realists and anti-realists, he claims, concerns the extent of the privilege assumed. According to Psillos, the Constructive Empiricist claims that scientists are privileged in generating theories that are *empirically adequate*, whereas the realist claims that scientists are privileged in generating theories that are *approximately true*.[2] Without the assumption of privilege, Psillos argues, there is little support for the Constructive Empiricist's claim that our best theories are empirically adequate.

---

[2] Contrary to what Psillos seems to imply, I think it is noteworthy that the sort of privilege that he says the Constructive Empiricist attributes to scientists, a tendency to develop theories that are empirically adequate, is more probable than the sort of privilege realists attribute to scientists. After all, there are likely to be more theories that are empirically adequate than theories that are apt to be approximately true.

Here, I think Psillos misrepresents the claim of the Constructive
Empiricist. Indeed, the Constructive Empiricist grants that our best
theories are superior to the ones with which they are compared. This
follows from what I referred to in the previous chapter as the Reliability Assumption. But the Constructive Empiricist is not committed
to the claim that our best theories are *in fact* empirically adequate.
What van Fraassen claims is that science *aims* for empirically
adequate theories (see van Fraassen 1980, 12). Hence, though scientists may evaluate competing theories with an eye to empirical
adequacy, from the fact that one theory is chosen over another, we
cannot infer that the chosen theory is empirically adequate. Scientists' proxy for either the truth (or the approximate truth) of a theory
or the empirical adequacy of a theory is the same. They look to see if
the theory can save the phenomena. That is, they look to see if it can
account for the observables. Being able to account for the observables, though, does not guarantee that the theory is true, or approximately true, or even empirically adequate. Indeed, a precise statement
of what van Fraassen means by empirical adequacy should make it
clear that a scientist cannot reasonably claim to be choosing an
empirically adequate theory when she chooses to accept a theory.
Van Fraassen claims that

a theory is empirically adequate exactly if what it says about the observable
things and events in the world is true — exactly if it "saves the phenomena" ...
Such a theory has at least one model that all the actual phenomena fit inside ...
This refers to *all* the phenomena; these are not exhausted by those actually
observed. (van Fraassen 1980, 12)

Thus, contrary to what Psillos suggests, the Constructive Empiricist
does not need to assume that scientists are privileged in their ability to
generate empirically adequate theories. All the Constructive Empiricist
needs to assume is that scientists are reliable in their judgments of the
comparative worth of competing theories. Hence, if there is an empirically adequate theory among the lot of theories from which a scientist is
choosing, given the Reliability Assumption, the scientist would be led
to choose the empirically adequate theory. Scientists, though, do aim to
develop theories that are *empirically successful*, where "empirically
successful" is understood to mean that they can account for the phenomena that have in fact been observed (but not the phenomena that
are potentially observable but have not yet been observed).

Thus, it seems that the indirect arguments for epistemic privilege do not provide us with compelling grounds for believing that scientists are in fact privileged. A number of realists, however, have developed *direct* arguments for epistemic privilege, including Psillos, Lipton, Trout, and Boyd. Let us consider these arguments. They turn out to be structurally similar to each other.

Psillos argues that our background theories, which are past successes, narrow the scope in our favor in our efforts to generate true theories (see Psillos 1999, 218–219; see also Ladyman 2002 and Trout 2016). That is, because scientists are guided by their background theories in developing new theories, they are inclined to hit upon either the true theory or a theory that is approximately true. The background theories constrain scientists' theorizing in a productive manner, keeping them focused on possible theories that are apt to be true. In this way, they are unlikely to be very wide of the mark in their aim to develop a true or approximately true theory.

A similar argument is presented by Lipton. He argues that "theory generation is highly constrained by background, and insofar as the background approximates the truth, we should not be surprised that our powers of generating true theories are substantially better than guesswork" (2004, 162). By "background," Lipton means background theories, in recognition of the fact that it is only in conjunction with various background theories that a hypothesis or theory implies specific observable phenomena. Lipton adds that "the striking successes of our best scientific theories actually provide empirical support for privilege; after all, success is much more likely with privilege than without it" (162–163).

Boyd (1985) offers a similar defense of realism. However, unlike Lipton and Psillos, Boyd believes that it is the reliability of their *methods* that ensures that scientists are epistemically privileged, and thus apt to be choosing from a set of theories that includes the true theory.

Trout is less sanguine than Boyd is about the efficacy of scientific methods in securing true or approximately true theories. He does not think that methodology itself can account for the success of our best theories. Rather, Trout insists that "scientific method only works well when you have a good enough background theory" (Trout 2016, 182). But Trout argues that, at least since Newton, and perhaps earlier, European scientists luckily hit upon theories that were at least approximately true. The theories he has in mind are the various corpuscular theories that were developed in the early modern period, the period that has traditionally been identified as *the* Scientific Revolution (see Trout 2016, 160). According to Trout, once scientists hit upon these approximately true theories, the methods they employ could aid them in refining these theories, developing theories that are even closer to the truth, as evidenced by the increasing accuracy with which scientists are able to make predictions. So on Trout's view, scientists' privilege was hard earned and has only been secured since the Scientific Revolution. Scientists now enjoy the sort of epistemic privilege that Psillos and Lipton attribute to them.

Privilege, in whatever form, these realists claim, significantly narrows the range of theories scientists consider. As a result, we should not be surprised when scientists hit upon a theory that is either true or approximately true.

But this realist line of reasoning is fallacious. Let us consider Psillos, Lipton, and Trout's argument first. No doubt, as Psillos, Lipton, and Trout claim, the background theories and assumptions that scientists work with will *narrow the range* of hypotheses or theories they are apt to develop. But it is far from clear that this supports the claim of epistemic privilege. In fact, it seems that background theories are as likely to be an impediment to developing a true theory as they are to be an aid to developing a true theory. For example, geologists working with the assumption that the continents are fixed are unlikely to entertain or develop hypotheses that ascribe motion to them. And if the continents do, in fact, move, such background assumptions will be an impediment. Similarly, physicists working with the assumption that all motion is due to contact between bodies are unlikely to develop a theory according to which there is action at a distance. And if there is, in fact, action at a distance, it will be an impediment to the advancement of science if scientists assume that all motion is due to contact. In fact, Newton's theory did meet with resistance from many continental physicists committed to the sort of contact physics developed by Descartes, then the dominant theory. Hence, even though, as Psillos, Lipton, and Trout argue, background theories will *narrow* the range of hypotheses scientists are apt to develop, they have not given us reason to believe that the net effect will be positive, inclining us toward developing true theories. Though background theories narrow scientists' thinking, the narrowing is not necessarily going to have a positive effect. Furthermore, we cannot assume that the narrowing does have a positive effect just because we have successful theories.[3]

Boyd's line of reasoning is subject to the same sort of objection that I have raised against Psillos, Lipton, and Trout. The methods scientists use will certainly restrict the range of theories they develop. And I concede that scientists' methods will surely aid them in excluding certain clearly false alternatives. But even granting that much, the realist must give us reason to believe that our current best scientific methods also ensure that scientists are apt to generate true or approximately true theories. Again, scientists' current methods may be an impediment to their aim of developing true theories. That is, scientists' commitment to current methods may prevent them from entertaining certain possibilities that would lead them closer to the truth.

This concern I raise against the direct arguments in support of epistemic privilege is similar to a concern Paul Feyerabend raises. Feyerabend notes that the accepted theory can sometimes be a serious impediment to scientific progress. As he notes, sometimes the data that may enable us to see that the accepted theory is false cannot even be detected without the aid of an alternative theory to bring the data into view. As Feyerabend explains, "*hypotheses contradicting well-confirmed theories give us evidence that cannot be obtained in any other way* " (Feyerabend 1988, 24; emphasis in original). So if scientists allow their current background theories to limit what hypotheses they entertain, they may be preventing themselves from detecting data that would undermine the accepted theory, and thus impede scientific progress. Similarly, the accepted methods in a field may also prevent scientists from gathering data that would lead them to reconsider the theories they accept. We cannot presume that the current methods are the final word on methodology.

Thus, it seems that the realists interested in defending the claim of epistemic privilege have yet to develop a compelling argument in support of epistemic privilege.

---

[3] Traditionally, realists have tried to infer the truth of our theories from their success. This is the core of the No Miracles Argument. Lipton and Psillos take this line of argument further by trying to infer the epistemic privilege of scientists from the success of our theories.

The Case for No-Privilege

In this section, I want to briefly consider some additional grounds for questioning the realists' claim that scientists are epistemically privileged. I will also clarify what the anti-realists' claim of no privilege entails.

Van Fraassen critically discusses epistemic privilege as part of his critical assessment of the realists' appeal to Inference to the Best Explanation. When one follows the inference rule of Inference to the Best Explanation, one moves from a "comparative judgment that this hypothesis is better than its actual rivals" to the conclusion that the better hypothesis is "more likely to be true than not" (see van Fraassen 1989, 143). It is that move, van Fraassen claims, that needs justification. In an effort to supply the justification, realists have sometimes appealed to our alleged epistemic privilege, according to which "we are by nature predisposed to hit on the right range of hypotheses" (van Fraassen 1989, 143). As a result, the truth is apt to be among the hypotheses we compare. Consequently, so the argument goes, it is reasonable to infer that the best of the lot we are choosing from is true or at least approximately true. Van Fraassen is unconvinced by this appeal to privilege. He doubts that there is any basis to support such a claim. Van Fraassen grants that there is nothing inconsistent in the view that we are epistemically privileged. But he also insists that the claim of epistemic privilege is not capable of justification (1989, 144).

Whereas van Fraassen suggests that the evidence in support of epistemic privilege is unconvincing, Mary Hesse (1976) explicitly defends a *no-privilege thesis*. She characterizes the no-privilege thesis in the following way: "our own scientific theories are held to be as much subject to radical change as past theories are seen to be" (1976, 264). The no-privilege thesis thus asks us to acknowledge the similarities between contemporary scientists and their predecessors. Hesse believes that the support for the thesis of no-privilege comes from an "induction from the history of science" (271).

Hesse argues that this no-privilege thesis supports skepticism about theoretical knowledge. According to Hesse,

every scientific system implies a conceptual classification of the world into an ontology of fundamental entities and properties ... But it is exactly these ontologies that are most subject to radical change throughout the history of science ... Therefore in the spirit of the principle of *no privilege* ... [our ontologies] must all be false. (Hesse 1976, 266; emphasis added)

Hesse is not a thoroughgoing skeptic. She even grants that some of our theoretical claims may be true (Hesse 1976, 266). But our theories, consisting of a conjunction of numerous theoretical claims, are most likely false.[4]

Despite her skepticism about theoretical knowledge, Hesse does not deny that scientific knowledge is growing. Rather, she believes that "there is accumulation of true observation sentences in the pragmatic sense that we have better learned to find our way about in the natural environment, and have a greater degree of predictive control over it" (1976, 274). Hesse emphasizes, however, that "this formulation of the growth of science does not presuppose privilege for our *theory*, because it is consistent with replacement of whole conceptual frameworks, including basic classifications and property assignments" (274, emphasis added). Hesse thus distinguishes between the instrumental growth of scientific knowledge, which is undeniable, and "a convergence of ontologies approximating better and better to a description of the true essence of the world" (275). With respect to instrumental growth, Hesse believes that scientists are making great progress. But we should not get too attached to our theories.[5]

The thesis of no-privilege thus involves the recognition that today's theories are as likely to be replaced in the future as were the successful theories of the past. It thus involves acknowledging a similarity between today's scientists and the scientists of the past. To claim that there is a profound asymmetry between today's scientists and the scientists of the past needs to be supported by evidence. But that is what realists have failed to supply. In Chapter 6, we will

---

[4] Hesse's intention is not merely to draw attention to the fact that a large conjunction, consisting of many claims, is likely to contain a false conjunct, and thus be false. Her concern is that theories are complex and that, given what we have seen throughout the history of science, the ontologies embodied in our theories are what ultimately prove to be wrong.

[5] It is worth noting that Mary Hesse was very sympathetic to Thomas Kuhn's *The Structure of Scientific Revolutions*, writing a very positive review of it in 1963 (see Hesse 1963).

look at recent attempts by realists to establish some sort of significant difference between today's scientists and the scientists of the past.

Reflections on epistemic privilege are not new. In the 1580s, Michel de Montaigne (1580/1948) questioned whether there are grounds for thinking that scientists are epistemically privileged. Montaigne was reflecting on this issue less than four decades after the publication of Copernicus' *De Revolutionibus*, and it seems clear that Montaigne recognized the importance of Copernicus' work. He explains that Copernicus believed "that it was the earth that moved, through the oblique circle of the Zodiac, turning about its axis" (Montaigne 1580/1948, 429). Montaigne further notes that "Copernicus has grounded this doctrine so well that he uses it very systematically for all astronomical deductions" (429). Hence, Montaigne was aware that the new theory could generate accurate predictions of the phenomena. In fact, Montaigne was quite perceptive, for, as we saw in Chapter 1, when he was writing, the vast majority of astronomers still believed that the Ptolemaic theory was correct (see Westman 1986/2003, 54).

Montaigne also recognized that neither of the two competing theories, that is, Ptolemy's theory and Copernicus' theory, would necessarily be the final word in astronomy. He asks rhetorically: "who knows whether a third opinion, a thousand years from now, will not overthrow the preceding two?" (1580/1948, 429). Again, Montaigne's remarks are prescient, given that less than a decade from when he published his remarks, Tycho Brahe published his theory, a theory that has been described as a hybrid of Ptolemy's and Copernicus' theories (see Brahe 1588/1970). And these three theories were not the only contenders vying for the attention and allegiance of early modern astronomers.

Montaigne generalizes from the case of Copernicus, recognizing that it is not a unique case in the history of science. As he explains:

when some new doctrine is offered to us, we have great occasion to distrust it, and to consider that before it was produced its opposite was in vogue; and, as it was overthrown by this one, there may arise in the future a third invention that will likewise smash the second. ... What special privileges [have the new theories we accept], that the course of our invention stops at them, and that to them belongs the possession of our belief for all time to come? They are no more exempt from being thrown out than were their predecessors. (1580/1948, 429)[6]

Thus, Montaigne recommends caution with respect to our contemporary theories. They seem as vulnerable to being discarded as were the theories they replaced. And, as Montaigne implies, those who accepted the older, rejected theories did so with the same conviction that many contemporary scientists now have for the theories they currently accept. We should not mistake the psychological feeling of certainty for evidence of the truth, likely truth, or approximate truth of our theories.

In summary, my aim here has been to show that arguments that try to explain the success of science by appealing to some sort of epistemic privilege have so far failed. Contrary to what some realists claim, scientists are not especially prone to develop theories that are true or approximately true. Neither the background theories nor the methods used by scientists provide adequate grounds for believing that scientists are especially prone to develop theories that are true or even approximately true. Consequently, realist arguments that depend upon a premise that asserts epistemic privilege are flawed.

---

[6] Ernst Mach thought that scientists could profit from learning the history of science. He argues that when scientists are instructed in the history of their discipline, they learn that theories that seem so compelling at one point come to be regarded as mistaken and are then discarded. Thus, they learn that mere *feelings* of certainty are often a poor guide to or indicator of theoretical truth. Mach's concern is that "whoever knows only one view ... does not believe that another has ever stood in its place, or that another will ever succeed it; he neither doubts nor tests" (Mach 1911, 17). Instruction in the history of science can thus foster a constructive modesty in scientists, and this can be a spur to further testing.

5 | *Four Pessimistic Inductions*

In the last chapter, we briefly examined appeals to the past record of science. Both Mary Hesse and Michel Montaigne suggested that, given the past record of science, it is unlikely that our contemporary theories will not be replaced by radically different theories in the future. In fact, this line of argument has figured prominently in the contemporary realism/anti-realism debate. The anti-realists' Pessimistic Induction (also called the Pessimistic Meta-induction) is frequently examined alongside the realists' No Miracles Argument (see Worrall 1989; Magnus and Callender 2004; Dicken 2016, for example). Some have suggested that the Pessimistic Induction is the anti-realists' strongest argument.

But there is quite a bit of controversy about the structure and aim of the Pessimistic Induction. A number of scholars have noted that there is more than one type of Pessimistic Induction in the philosophical literature (see, for example, Lewis 2001, 371; Lange 2002, 281; Mizrahi 2013; Ruhmkorff 2013, 410). Juha Saatsi (2005), for example, argues for the importance of distinguishing between "Laudan's argument and Putnam's rhetoric" (see Saatsi 2005, 1091). He also believes that Henri Poincaré presents a Pessimistic Induction (see Saatsi 2005, 1088).

In this chapter, I want to review and evaluate four different Pessimistic Inductions. As we will see, philosophers of science have appealed to Pessimistic Inductions for a variety of reasons. Though the Pessimistic Induction is commonly, perhaps exclusively, thought of as an anti-realist argument, even some realists have appealed to some form of Pessimistic Induction. My aim is to advance our understanding of (i) what the various Pessimistic Inductions can teach us about science and (ii) the threat posed to scientific realism by the various Pessimistic Inductions. I also aim to indicate how one could strengthen these arguments.

Before proceeding, it is worth articulating the basic general form of a Pessimistic Induction. A Pessimistic Induction is an inductive argument that draws a conclusion from the rejection of many successful scientific theories in the past. Sometimes a conclusion is drawn about the prospects of the theories that are currently accepted, and sometimes an inference is drawn about the prospects of future theories, those not yet developed or entertained by scientists. I believe the way I have characterized Pessimistic Inductions is not contentious, and similar characterizations can be found even in the writings of realists. Anjan Chakravartty, for example, characterizes the Pessimistic Induction in the following way:

> PI can ... be described as a two-step worry. First, there is an assertion to the effect that the history of science contains an impressive graveyard of theories that were previously believed [to be true], but subsequently judged to be false ... Second, there is an induction on the basis of this assertion, whose conclusion is that current theories are likely future occupants of the same graveyard. (Chakravartty 2008, 152)

Let us now consider the various versions of the Pessimistic Induction in detail.

Putnam's Pessimistic Meta-Induction

Hilary Putnam presents the clearest example of the Pessimistic Induction advanced as an argument against scientific realism. In laying out the argument, Putnam discusses cases from the history of science where the alleged entities referred to by theoretical terms either (i) turned out not to have the properties they were assumed to have or (ii) turned out not to exist at all. Bohr's electron is an example of the former type of case, and phlogiston is an example of the latter (see Putnam 1978, 24). Our current best theory of the electron suggests that it does not have some of the properties Bohr ascribed to it.[1] For example, *"in the Bohr model* [of the hydrogen atom] *... it is not possible for the orbital angular momentum to be zero ...* [and] if the electrons have the same energy *... they cannot have different values for the orbital angular momentum"* (Cutnell and Johnson 2001, 925; emphasis in original).

---

[1] This remark is a bit misleading, implying that there is a single best theory of the electron. Ian Hacking notes that "even people in a [research] team, working on different parts of the same large experiment, may hold different and mutually incompatible accounts of electrons" (Hacking 1983, 264). Hence, there may not be a *single* best theory of the electron.

In contrast, in quantum mechanics "the orbital angular momentum may be zero ... [and] the electrons could have different angular momenta, even though they have the same energy" (925). Thus, Bohr's conception of the electron misrepresents it in a number of ways. And the concept "phlogiston" has no place in modern chemistry. "Phlogiston" was, in one sense, replaced by "oxygen," though the types of substances designated by these terms have very different properties. Perhaps the most fundamental difference is that phlogiston was deemed to be a substance that is emitted into the atmosphere by burning substances, whereas oxygen is thought to be a substance that is taken out of the atmosphere when a substance burns (see Henry 2012, 170). Only the most Whig historians of science could claim that oxygen is just phlogiston by another name.

In light of these cases, Putnam raises the following question: "what if *all* the theoretical entities postulated by one generation (molecules, genes, etc., as well as electrons) invariably 'don't exist' from the standpoint of later science?" (Putnam 1978, 24; emphasis in original). If we do find that this is the case, that our theoretical postulates *invariably* turn out not to refer, Putnam claims that

the following meta-induction becomes overwhelmingly compelling: *just as no term used in science of more than fifty* (or whatever) *years ago referred, so it will turn out that no term used now* (except maybe observation terms ...) *refers*. (Putnam 1978, 25; emphasis in original)

That is, the history of science, filled with scientists' failed attempts to develop theories with theoretical terms that genuinely refer, seems to support the conclusion that today's best theories will meet a similar fate. The central theoretical terms of our current theories will also be discovered to not refer.[2]

There are three important points to note about this specific formulation of the Pessimistic Induction. First, Putnam does not actually claim

[2] This is the version of the Pessimistic Induction that Robert Nola criticizes (see Nola 2008). In his attack on it, Nola appeals to the causal theory of reference, a theory that Putnam is partly responsible for developing. Nola also suggests that an Optimistic Induction is better supported than Putnam's Pessimistic Induction from the history of science. Michael Devitt (2011), though, rightly notes that it is problematic to attempt to address the Pessimistic Induction by appealing to a theory of reference, given the disagreement among philosophers about theories of reference (289). Incidentally, Devitt is a realist; his concern is that this appeal to a specific theory of reference does not provide the strongest defense of realism.

that our theoretical postulates inevitably turn out not to refer. He merely asks us to consider the possibility. Thus, Putnam is not suggesting that we have the evidence to support the skeptical conclusion. Indeed, he does not endorse the argument. It is presented as part of a discussion of the strongest arguments for and against scientific realism.[3]

Second, Putnam's formulation of the argument makes a universal claim. It suggests that we may find that "*all* the theoretical entities postulated by one generation ... invariably 'don't exist.'" I think that even the most skeptical contemporary anti-realists would think that it is doubtful that *all* the theoretical entities postulated more than fifty years ago do not exist. In fact, it is probably doubtful that all the theoretical entities postulated even two hundred years ago do not exist. But the anti-realist need not make such a grand claim with such a broad scope. And clearly a more plausible argument could be advanced that does not make the *universal* claim. Provided many of the theoretical entities postulated turn out not to exist, there are grounds for skepticism about the entities postulated by our current theories. How many of our past successful theories would need to be shown to contain theoretical terms that are not genuinely referring in order to support a skeptical conclusion is unclear.

Larry Laudan famously suggested that "for every highly successful theory in the past of science which we now believe to be a genuinely referring theory, one could find half a dozen once successful theories which we now regard as substantially non-referring" (see Laudan 1981, 35). This estimation has become the focus of some debate (see, for example, Lewis 2001, 375; Wray 2013). But we should not attach too much weight to Laudan's estimation. Less than ten pages before this remark in "Confutation of Convergent Realism," Laudan explicitly acknowledges the need for a systematic collection of data relevant to testing the hypothesis that past successful theories are "ones whose central terms genuinely refer" (see Laudan 1981, 26). He notes that "a proper empirical test of this hypothesis would require extensive sifting of the historical record of a kind that is not possible to perform here" (Laudan 1981, 26).[4]

---

[3] Putnam suggests that the No Miracles Argument is the strongest argument in support of realism. I discuss the No Miracles Argument in detail in Chapter 9.

[4] Laudan was involved in a large-scale project to test the claims philosophers of science make about science (see Donovan, Laudan, and Laudan 1988).

We would benefit from a systematic collection of evidence in order to determine whether or not a strong *inductive* argument is supported. Indeed, it is often taken for granted that such evidence is readily available. Anyone familiar with the history of science can readily think of some theories that were quite successful that have subsequently been replaced by better theories. But a strong inductive argument would need an evidential base that is more systemically collected – and the evidence needs to be drawn from cases that are similar in the relevant respects, for example, taken from a period of time in which scientists were employing methods not unlike the methods used by contemporary scientists. Recently, Moti Mizrahi has attempted to test the inductive version of the Pessimistic Induction more systematically (see Mizrahi 2013). I will discuss Mizrahi's efforts in this regard in Chapter 7.

Third, the plausibility of Putnam's version of the Pessimistic Induction depends crucially on the period in the history of science from which the inductive inference is drawn. Putnam rather unreflectively suggests that theories developed more than fifty years ago may provide an appropriate inductive base. There are two concerns with this suggestion. First, unless one goes back far enough in the history of science, one might not be able to construct a compelling and well-supported inductive argument. That is, if one focuses on the recent history of science, one may find that most theories have not in fact been rejected on the grounds that they contain theoretical terms that (we now believe) do not refer. For example, we might find that only about 30 percent of the theories developed between 1860 and 1960 have been rejected.[5] If this were the case, the skeptical conclusion would be undermined. A second concern is that the plausibility of the Pessimistic Induction may depend upon grouping together theories developed in radically different times, and under significantly different circumstances. There are a number of different versions of this concern in the scholarly literature raised by realists criticizing the Pessimistic Induction.

One variation focuses on developments in methodology. Richard Boyd (1983; 1985), Sherrilyn Roush (2010, 55), and Michael Devitt

---

[5] Ludwig Fahrbach (2011) makes a conjecture of this sort, though he is not so specific as to claim that 30 percent is the appropriate number. He argues that the vast majority of theories have been developed in the recent history of science, since 1900. I discuss Fahrbach's view in detail in the next chapter.

(2011) have argued that developments in methodology pose a serious challenge to the anti-realist's Pessimistic Induction. Their concern is that if we go too far back in the history of science, we may be grouping together (i) theories developed before scientists had made some significant developments in methodology with (ii) more recently developed theories that were developed with the aid of superior methods. These various theories may not form a uniform group from which it is reasonable to draw a well-supported inductive inference about the fate of today's best theories. Maybe only the fate of our most recently developed theories is relevant to determining what we can expect of today's best theories. We will look at this argument in more detail in the next chapter. Specifically, I will examine Devitt's attempt to block the Pessimistic Induction by appealing to developments in methodology.

Marc Lange (2002) also argues that it is illegitimate to construct an inductive argument by drawing an inference from *all* past rejected theories. His concern, though, is somewhat different from the concern raised by Boyd, Roush, and Devitt. Lange suggests that we are likely to find that in some scientific specialties there is quite a rapid turnover rate, with new theories taking the place of older theories in relatively quick succession, whereas in other scientific specialties theories may have more staying power, and thus a lower turnover rate. In a specialty where only two theories have been rejected in the past five hundred years, each replaced by a better one, one may rightly be reticent to draw a skeptical conclusion about the currently accepted theory. In contrast, in a specialty where there has been rapid turnover, there is more reason for skepticism about the currently accepted theory. For example, there were at least three different theories of light, one succeeding the other, in less than 120 years between the early 1800s and early 1900s (see Fahrbach 2011, 142, figure 1, for a vivid depiction of this).[6] In contrast, the Ptolemaic theory enjoyed a period of about 1,300 years virtually unchallenged, until the mid-1500s. The

---

[6] Fresnel's wave theory replaced the particle theory developed earlier by Newton. Fresnel's theory was replaced by Maxwell's theory, which dispensed with the ether, the medium through which Fresnel believed light waves traveled. Maxwell regarded light as a periodic disturbance "in the 'disembodied' electromagnetic field" (see Worrall 1989, 116). "Einstein reintroduced particles; and finally the 'probability waves' of Quantum Mechanics came up" (Fahrbach 2011, 141–142).

former type of case is more apt than the latter to lead to skepticism about theoretical knowledge.

Interestingly, Lange is suggesting that the scientific specialty may be the right unit of analysis when considering the Pessimistic Induction. A more global argument that groups together theories drawn from many different scientific specialties may be untenable. Incidentally, Magnus and Callender also suggest that wholesale arguments for or against realism, "arguments about *all* or *most* of the entities posited in our best scientific theories," lead to ennui and endless and irresolvable debate (2004, 321; see also Dicken 2016, chapter 5). Magnus and Callender, though, suggest that the proper unit of analysis is not the scientific field, but rather the theoretical entity.

These concerns alert us to the fact that the anti-realist appealing to the Pessimistic Induction needs to steer between the following two potential threats to a strong inductive argument. On the one hand, the anti-realist needs as large an induction base as possible to have a well-supported argument. This will lead him to look far back into the history of science. On the other hand, the anti-realist needs to be careful not to indiscriminately group together theories developed under very different circumstances, and thus generalize from an unnatural grouping. This will lead him to look to the history of science in a more restricted way, drawing data only from those periods in which the practice of science has been more or less continuous with contemporary practices.

The Pessimistic Induction as a *Reductio Ad Absurdum*

There is another variation of the Pessimistic Induction common in the philosophical literature, a reconstruction by scientific realists set on criticizing the argument. These realists reconstruct the Pessimistic Induction as a *reductio ad absurdum*. The argument is reconstructed in the following manner:

P1. Assume that "currently successful theories are approximately true."

P2. "If currently successful theories are truth-like, then past theories *cannot* have been."

P3. "These ... false theories [of the past] were, nonetheless, empirically successful."

C.   Therefore, "empirical success is not connected with truth-likeness and truth-likeness cannot explain success." Therefore, "the realist's potential warrant for [the claim that currently successful theories are approximately true] is defeated." (Psillos 1999, 102)

It is worth noting that this version of the Pessimistic Induction is concerned with the truth or falsity of successful theories, whereas the version developed by Putnam is concerned with whether or not successful theories have genuinely referring theoretical terms. Clearly, though, as Laudan notes, if a theory has central theoretical terms that do not genuinely refer, it will not be true.[7]

It is ironic that this version of the Pessimistic Induction has played such a central role in the current debate between realists and antirealists. After all, this version of the Pessimistic *Induction* is not an inductive argument at all. It is a deductive argument, as are all *reductio ad absurdum* arguments. It is also ironic that this particular formulation of the Pessimistic Induction is so often attributed to Laudan (see Psillos 1999, 102; Lewis 2001, 373; Saatsi 2005, 1088–1089; Devitt 2011; Dellsén, forthcoming). This argument is alleged to be found in Laudan's "Confutation of Convergent Realism" (see Laudan 1981).[8] But in fact Laudan does not present this argument in that paper. His argumentative strategy in that paper is *deductive*. Timothy Lyons characterizes Laudan's argument as a *meta-modus tollens* (see Lyons 2002).

Laudan's aim is to show that key realist claims about the connection between (i) scientific success and theoretical truth and (ii) scientific success and genuine reference are false. Specifically, he aims to show that a theory's having genuinely referring theoretical terms is neither

---

[7] Surprisingly, Hardin and Rosenberg argue against Laudan's claim that "'the *realist would never want to say that a theory was approximately true if its central theoretical terms failed to refer*'" (Laudan, cited in Hardin and Rosenberg 1982, 606). They argue that "Mendel's 1866 theory, embodying laws of segregation and assortment, clearly constitutes the first in a sequence of successive theories which are held by life scientists to constitute a series converging on the truth" (Hardin and Rosenberg 1982, 606). Kitcher holds a similar view (see Kitcher 1993, 137).

[8] Kitcher (1993, 136, Note 13), Forster and Sober (1994, 28), Papineau (1996, 14), Lipton (2004, 145), Magnus and Callender (2004), Chakravartty (2007), Fahrbach (2011, 141), and Chang (2012, 270) attribute the Pessimistic Induction to Laudan, but they do not reconstruct it as a *reductio ad absurdum*. Greg Frost Arnold (2011) is more cautious, merely noting that Laudan's "Confutation" is responsible for the recent attention on the Pessimistic Induction.

necessary nor sufficient for the theory to be successful, and that a theory's being true (or approximately true) is neither necessary nor sufficient for the theory to be successful. Thus, Laudan is concerned with *both* the link between truth and success *and* the link between genuine reference and success, contrary to what Psillos's reconstruction of the Pessimistic Induction suggests.

As far as Laudan is concerned, a single successful theory that is false would falsify the realist's claim that (all) successful theories are true, and a single successful theory that refers to a nonexistent type of entity would falsify the realist's claim that (all) successful theories have genuinely referring theoretical terms. After all, success cannot be a necessary condition for the approximate truth of a theory if there is even one approximately true theory that is not successful. And success cannot be a sufficient condition for the approximate truth of a theory if there is even one successful but false theory. A careful examination of Laudan's arguments in "Confutation" shows that he relies on just a few examples to undermine the realists' claims about the alleged connections between (i) truth and success and (ii) genuine reference and success. Thus, a catalogue of failures, the sort of evidence we might expect in an inductive argument, is unnecessary. In this respect, the famous list of failed theories that Laudan does provide in "Confutation" is unnecessary and constitutes overkill.[9]

This version of the Pessimistic Induction has taken on a life of its own, and it is discussed widely by scientific realists and anti-realists. It is presented by Psillos, for example, and discussed by Devitt, Peter Lewis, and others (see Psillos 1996, S307, and 1999, 102–103; Lewis 2001, 373; Lange 2002, 282; Devitt 2011). And a number of anti-realists have responded to criticisms of the Pessimistic Induction by explicitly addressing this version of the argument. Saatsi, for example, explicitly criticizes Lewis's attack of this formulation of the Pessimistic Induction (see Saatsi 2005, 1089).

Psillos's key criticism of this version of the Pessimistic Induction targets Premise 2 above, the claim that if currently successful theories are truth-like, then past theories *cannot* have been (see Psillos 1999).

---

[9] Laudan's list has generated a vast body of critical literature (see, for example, Bishop 2003, § 2.2; Mizrahi 2013, 3219–3220). Moti Mizrahi is especially concerned that Laudan's examples are not a random sample of the target population of successful theories.

Psillos's *divide et impera* strategy is meant to show that one can acknowledge the failings of past theories as measured against currently accepted theories, yet still insist that those past theories had truth-like elements or features. In fact, Psillos insists that these truth-like features of the now-rejected theories (i) are the features that are retained in currently accepted theories and (ii) are the features responsible for the successes of the now-rejected theories (see Psillos 1996, especially § 2).

I think that Psillos and others are correct to claim that this is not the strongest argument from the history of science in support of anti-realism. But I do not think that it does justice to the insight that Laudan's "Confutation of Convergent Realism" offers to the realism/anti-realism debate. Laudan's paper is better seen as a contribution to our understanding of the role of the theoretical virtues in theory evaluation. I will analyze his argument in "Confutation of Convergent Realism" in detail in Chapter 8.

Realism and the Pessimistic Induction

The Pessimistic Induction is not always employed as an attack on realism. Sometimes realists appeal to the argument in order to clarify some fact about science. More precisely, they appeal to the Pessimistic Induction in order to show that some commonly held view about science is mistaken, and the success of science is something quite different, perhaps more circumscribed, than many people think. Henri Poincaré, John Worrall, and Nicholas Rescher appeal to the Pessimistic Induction for this purpose.

Consider Poincaré's appeal to the Pessimistic Induction. Reflecting on the history of science in *Science and Hypothesis*, Poincaré claims that

the ephemeral nature of scientific theories takes by surprise the man of the world. Their brief period of prosperity ended, he sees them abandoned one after another, he sees ruins piled upon ruins; he predicts that the theories in fashion to-day will in a short time succumb in their turn, and he concludes that they are absolutely in vain. This is what he calls the *bankruptcy of science*. (Poincaré 1905/2001, 122; emphasis in original)

Contrary to what some seem to suggest, Poincaré does not really think that science is bankrupt (see, for example, Saatsi 2005, 1088). Rather, he is careful to distinguish between the common person's view of science and the reflective scientist's view. Poincaré insists that the common person's "scepticism is superficial." The common person is led to think that theorizing is done in vain (Poincaré 1905/2001, 122). But this is not Poincaré's view. He believes that theorizing plays an indispensable role in science.

In *The Value of Science*, Poincaré is more explicit about the positive insight he draws from his version of the Pessimistic Induction. Here, as before, he notes that "at first blush it seems to us that ... theories last only a day and that ruins upon ruins accumulate" (Poincaré 1913/2001, 348). But, again, Poincaré insists that first impressions are misleading. He notes that "if we look more closely, we see that what thus succumb are the theories ... which *pretend to teach us what things are*" (Poincaré 1913/2001, 348–349; emphasis added). Thus, he wants us to see that there is a limit to what we can expect to learn from our scientific theories. Theories cannot teach us "what things are." But he insists that theories, even theories we have discarded, often contribute to the progress of science. As Poincaré notes,

there is in [the discarded theories] something which usually survives. If one of them has taught us a true relation, this relation is definitively acquired, and it will be found again under a new disguise in the other theories which will successively come to reign in place of the old. (Poincaré 1913/2001, 349)

Thus, Poincaré invokes the Pessimistic Induction in order to show how misguided the common person's view of science is, and to explain where the real success that scientists achieve lies. As far as he is concerned, the real advances in science are made with respect to our understanding of the structure of reality, that is, the genuine relations captured in the formulas that persist through changes of theory.[10]

---

[10] Whether Poincaré is aptly characterized as a realist or not is open to debate. As we will see in a moment, Worrall has co-opted Poincaré as the founder of Structural Realism. Recently, though, Milena Ivanova (2015) has questioned whether Poincaré is a *realist*. According to Ivanova, Poincaré, like Kant, believes that "we cannot discover facts about the world ... independent of our cognitive apparatus" (2015, 87). She takes this claim to conflict with the realists' claim that "there is a mind-independent reality that scientific theories discover" (Ivanova 2015, 87). Karl Popper (1935/2002) treats Poincaré as a conventionalist, and thus an opponent of realism. I have discussed Popper's assessment of Poincaré's conventionalism elsewhere (see Wray 2015b).

This same line of argument is developed by Worrall in his defense of Structural Realism, a position inspired by Poincaré's view. According to Worrall,

it is ... logically possible that although all previous theories were false, our current theories happen to be true. But to believe that we have good grounds to think that this possibility may be actualised is surely an act of desperation — it seems difficult indeed to supply any halfway convincing reason to hold that we can legitimately ignore the possibility that the future history of science will be similar to the past history of science and therefore to ignore the possibility that our current theories will eventually be replaced in the way that they themselves replaced their predecessors. (2007, 129)

Like anti-realists, Worrall is skeptical that our current best theories accurately describe the unobservable *entities* they purport to describe. And he also believes that it is "an act of desperation" to think that contemporary scientists have transcended the difficulties that earlier scientists faced. But, unlike anti-realists, Worrall believes that there is good evidence for the claim that scientists are able to get at the underlying structure of reality. Our success in this respect, he claims, is evidenced by the fact that mathematical formulas developed in one theory are sometimes retained by the successor theory (see Worrall 1989; 2007). Most frequently mentioned in support of this claim is the fact that Maxwell was able to retain Fresnel's equations even though he had a radically different conception of light. The resilience of the equations – their persistence through radical theory change – suggests that they must be latching on to some aspect of reality. The Pessimistic Induction, Worrall suggests, helps us see that progress in science is not a function of scientists converging on and refining an ontology that reflects the basic entities in the world. Instead, it involves developing our understanding of the structure of reality, a feature of reality captured by the mathematical formulas scientists have developed.

Rescher also appeals to the Pessimistic Induction, though he does not support anti-realism. Rescher wants us to see that a key claim commonly associated with science is mistaken; specifically, that progress involves convergence. Rescher argues that "historical experience shows that there is every reason to expect that our ideas about nature are subject to radical changes as we 'explore' parametric space more extensively" (Rescher 1987, 15). He believes that, over time, scientists are gathering data on more and more variables. This is what Rescher means by exploring parametric space more extensively. But sometimes scientists discover that they are unable to reconcile the new data they collect with the accepted theory. This is what leads them to develop a new theory. This process, Rescher claims, is driven to a large extent by innovations in instrumentation in science. Rescher argues that "the technologically mediated entry into new regions of parametric space constantly destabilizes the attained equilibrium between data and theory" (Rescher 1987, 15). Importantly, Rescher is suggesting that as long as science is developing, one should expect theories to be replaced, even our best contemporary theories.

It is at this point that Rescher draws his pessimistic conclusion. He claims that "the history of science is a history of episodes of leaping to the wrong conclusion" (Rescher 1987, 16). He believes that scientists tend to over-generalize when they draw inferences from the data they have. And as they explore hitherto unexplored "parametric space," they often discover that the generalizations they or their predecessors made are false. A clear example of this is the trouble Newton's theory faced when scientists began studying particles moving at speeds close to the speed of light. It was discovered that Newton's laws had a more circumscribed application than Newton and his contemporaries thought.

Rescher, though, does not take this failure on the part of scientists as grounds for accepting some form of anti-realism. Instead, he suggests that the history of science, as described above, undermines convergentism, a particular theory of scientific progress. Convergentism is the view that successive theories in a series of changes of theory bring us ever closer to the truth. Rescher insists that "convergentism ... lacks support [from] ... the history of science" (Rescher 1987, 24–25).[11]

Again, like Poincaré and Worrall, Rescher believes that the Pessimistic Induction helps us better understand the real success in science. Clearly, as scientists explore new regions of parametric space, their knowledge of the world increases. The notion of convergence,

---

[11] Laudan devotes a section of "Confutation of Convergent Realism" to criticizing realists' appeals to convergence (see Laudan 1981, § 6). He argues that "some of the most important theoretical innovations have been due to a willingness of scientists to violate the cumulationist or retentionist constraint which realists enjoin 'mature' scientists to follow" (39).

however, is not insightful in this context. In fact, Rescher (1978) suggests that scientific progress is characterized by the proliferation of scientific specialties (see Rescher 1978, 229, table 3).[12]

It is worth briefly commenting on why these arguments warrant being called Pessimistic Inductions. Recall the characterization with which I began this chapter. The arguments discussed in this section are built on a consideration of the many once successful but now discarded theories. They draw a conclusion about the prospects of contemporary successful theories. Neither Poincaré, nor Worrall, nor Rescher believes that we have good reason to think that today's theories will escape the fate of past successful theories. What makes their use of the Pessimistic Induction different from anti-realists' appeals to the Pessimistic Induction is that, unlike the anti-realists, they believe that science is making significant progress that warrants endorsing some form of *realism*. As realists, they appeal to the Pessimistic Induction in order to clarify what scientists can and cannot achieve. Reflecting on the Pessimistic Induction, Poincaré, Worrall, and Rescher were led to develop modest forms of realism.

Incidentally, Gerald Doppelt (2005) also believes that the Pessimistic Induction helps clarify what a plausible form of realism entails or commits one to. In fact, there are now a number of modest forms of realism. Entity Realism, for example, involves a commitment to the belief in the existence of the theoretical entities scientists manipulate routinely in laboratory operations, but recognizes that scientists may be mistaken about some of the properties that they ascribe to those entities (see Hacking 1983, chapter 16). Alternatively, Anjan Chakravartty has suggested that the most viable form of realism is "realism about well-confirmed *properties*" rather than entities (see Chakravartty 2008, 155). Even Psillos's *divide et imperia* strategy constitutes a concession to the anti-realist. All these realists take seriously the fact that theory change is ubiquitous in the history of science. Their chief point of difference with anti-realists is that they believe there is some form of continuity through changes of theory that provides warrant for realism.

---

[12] Thomas Kuhn (1991/2000) makes a similar remark about the increasing specialization that characterizes science (97–98). For an extended discussion and defense of Kuhn's views on specialization, see Wray (2011, chapter 7).

Stanford's *New* Pessimistic Induction

The contemporary realism/anti-realism debate was recently reinvigor-
ated and set in a new direction with the development of Kyle Stanford's
new Pessimistic Induction. Stanford refers to his version of the Pessimistic Induction as the Argument from Unconceived Alternatives (see
Stanford 2001; see also Stanford 2006). Unlike the philosophers discussed in the previous section, Stanford is interested in defending an
anti-realist position.

This argument does not start from a consideration of the many once
successful and now rejected theories. Rather, it looks at the future of
science from the perspective of the scientists who worked in the past
with theories that are no longer accepted. These theories that are no
longer accepted are assumed to be false. And rather than emphasizing
the failures of the past, Stanford's argument focuses on the superiority
of theories developed more recently.

The Argument from Unconceived Alternatives notes that, almost
invariably in any scientific field, theories developed more recently are
able to account for the same data that their predecessors could account
for, and more (Stanford 2001, S9). Further, these successor theories
were unconceived at the time when their predecessors were the
accepted theories. On the basis of these reflections, Stanford proposes

the following New Induction over the History of Science: that we have,
throughout the history of scientific inquiry and in virtually every field,
repeatedly occupied an epistemic position in which we could conceive of
only one or a few theories that were well-confirmed by the available evidence, while subsequent history of inquiry has routinely (if not invariably)
revealed further, radically distinct alternatives as well-confirmed by the
previously available evidence as those we were inclined to accept on the
strength of that evidence. (Stanford 2001, S9)

Stanford thus suggests that we have good inductive grounds for believing that today's best theories are likely to be replaced sometime in the
future by hitherto unconceived alternative theories.

Stanford's argument, unlike the *reductio ad absurdum* attributed
to Laudan, is a genuine inductive argument.[13] Stanford supplies a list

---

[13] Recently, Timothy Lyons reconstructed Stanford's argument as a *deductive*
argument, specifically a *modus ponens* argument, but one in which some of the
premises are "inductively grounded" (see Lyons 2013, 372).

of theories that support the inductive inference. He discusses, for example, the sequence of theories in physics from Aristotle's theory to the mechanistic theory associated with Descartes to Newtonian mechanics, and finally to the currently accepted theory developed by Einstein (see Stanford 2001, S9). Stanford suggests that every field is like physics, where earlier theories are replaced by new theories that make substantially different assumptions about the nature of reality, and he provides a list of examples from a variety of scientific specialties to support his claim (see Stanford 2001, S9). These more recently developed theories were unconceived by earlier scientists.

Earlier, I mentioned that realists had raised the criticism that Putnam's formulation of the Pessimistic Induction fails to acknowledge the progress that has been made in science, specifically in instrumentation and methodology. A key strength of Stanford's Pessimistic Induction is that it seems to acknowledge that scientists in the more recent past were probably working with better methods and more precise instruments than their predecessors. In this sense, he is making a concession to the realist. Given the developments in methodology and instrumentation, it is not surprising that today's theories are superior to the theories they replaced. Indeed, that is what one would expect. But Stanford does not take this as grounds for believing that scientists are getting ever closer to the truth. Rather, he argues that changes of theory often involve radical changes in ontology that undermine the plausibility of convergentism.

When Stanford presents the New Pessimistic Induction in *Exceeding our Grasp*, he draws attention to a different dimension of this particular Pessimistic Induction. He claims that his argument, unlike the traditional Pessimistic Induction, "concerns *theorists* rather than *theories* of past and present science" (Stanford 2006, 44). Elaborating, Stanford claims that

the problem of unconceived alternatives and the new induction suggest that ... present *theorists* are no better able to exhaust the space of serious, well-confirmed possible theoretical explanations of the phenomena than past theorists have turned out to be. (Stanford 2006, 44; emphasis added)

How important this shift from theories to theorists is to the debate between realists and anti-realists has yet to be determined. Stanford does not actually deploy the distinction all that much in his attack on realism. Even critics who acknowledge Stanford's focus on theorists make little of it in their critical analysis (see, for example, Chakravartty 2008, 150–151).[14]

Stanford's new Pessimistic Induction has also been subjected to critical scrutiny. Chakravartty (2008), Devitt (2011), and P. D. Magnus (2006), for example, question whether Stanford's argument from unconceived alternatives poses a threat different from the threat posed by the original Pessimistic Induction (see Chakravartty 2008, 149). Chakravartty claims that "a successful response to the Pessimistic Induction would likewise defuse the problem of unconceived alternatives" (149). Like the original Pessimistic Induction, Stanford's New Induction suggests that theory change is ubiquitous in the history of science. Chakravartty, though, is concerned that "the real question of interest ... is whether there is anything like a principled continuity across scientific theories" (Chakravartty 2008, 153). Chakravartty does not believe that the New Pessimistic Induction raises the level of threat on this question. Provided there is some sort of "principled" continuity through changes of theory, some form of realism is defensible. And Chakravartty believes that "there appears to be a great deal of preservation of mathematical structure across theories over time" (Chakravartty 2008, 155). The preservation of mathematical structure through changes of theory is thus the basis for a plausible and defensible form of realism, Chakravartty claims.

Devitt (2011) also thinks that the New Pessimistic Induction is vulnerable to the same sort of criticisms that have threatened the traditional Pessimism Induction. But Devitt specifically argues that developments in methodology undermine both Pessimistic Inductions. We will look at Devitt's argument briefly in the next chapter. And Magnus argues that "Stanford's New Induction ... merely recapitulates familiar philosophical conundra" (Magnus 2006, 295).

It is worth bringing some order to this analysis of the various Pessimistic Inductions from the history of science. As we have seen above, some think that the Pessimistic Induction need not pose a threat to scientific progress. Some realists appeal to Pessimistic Inductions to clarify where the real success of science lies. Many acknowledge

---

[14] Lyons is an exception. He believes that Stanford's shift from theories to theorists is significant (see Lyons 2013, 371–372).

that revolutionary changes of theory are ubiquitous in the history of science, and that such changes are likely to continue to occur in the future. Even some realists recognize this (see, for example, Worrall 1989). Despite the recurring changes of theory, though, scientists do seem to have a richer understanding of the world. As Rescher notes, today's scientists are theorizing about phenomena that were unaccounted for by scientists in previous generations. So the Pessimistic Induction does not necessarily spell doom for scientific realism, at least not all forms of it. But given the history of science, realists need to be cautious. Those who develop and defend the various modest forms of realism tacitly acknowledge this. And the Pessimistic Induction is consistent with progress in science, at least with respect to scientists' knowledge of observables. But, importantly, the anti-realists advancing these arguments never intended to question the progress of science. Their concern is with respect to the realists' claims about the growth of *theoretical knowledge.*

All of the Pessimistic Inductions, it seems, aim to show that the differences between contemporary scientists and scientists of the past are, in some important sense, negligible. The point is not to suggest that contemporary scientists do not have knowledge of many things that their predecessors did not know about. Clearly they do. Rather, the point is that scientists today face similar challenges and barriers in their efforts to know the things they are investigating at the research frontier to those faced by their predecessors in their investigations at the research frontier. In this respect, all the Pessimistic Inductions are based on a premise of No Privilege. In the next chapter, I want to examine some recent attempts by realists to show that (i) contemporary scientists are epistemically privileged and (ii) this privilege undermines the Pessimistic Induction.

Before moving on, it is worth noting that not all anti-realists appeal to the Pessimistic Induction. Bas van Fraassen (2007), one of the philosophers most responsible for the renewed interest in anti-realism in the last four decades, explicitly distances himself from the Pessimistic Induction. He claims that he is "quite proud never to have relied on the so-called Pessimistic Induction ... any more than on [the] Argument from Underdetermination" (van Fraassen 2007, 347). Van Fraassen puts more stock in the Argument from Underconsideration, discussed earlier. In Chapters 9 and 10, we will look in detail at other sorts of considerations that motivate van Fraassen's anti-realism.

I am inclined to agree with van Fraassen, to some extent. The challenge that the history of science raises for realists is not best understood in terms of a Pessimistic Induction from past failures. Were this the principal challenge for realism, it seems that a detailed and systematic survey of the history of science would be necessary. I think the realist should feel threatened by even one radical change of theory in a scientific field. I will discuss this issue further in Chapter 7.

6 | Pessimism, Optimism, and the Exponential Growth of Science

In this chapter, I want to bring together topics that I discussed at length in the previous two chapters, the Pessimistic Induction and the notion of epistemic privilege. It is worth clarifying how they are related. The various Pessimistic Inductions, insofar as they seek to generalize from past theories to future theories, emphasize the similarity between today's scientists and scientists of the past. One strategy that some realists have used in their attempts to undermine or block any sort of Pessimistic Induction is to emphasize the differences between today's scientists and scientists of the past. This strategy is, in a sense, yet another sort of appeal to epistemic privilege. In this case, the privilege is alleged to have been acquired relatively recently, perhaps only in the last century or so. Past scientists may have been groping in the dark, but, so the argument goes, we have now reached a new age where scientists are epistemically privileged. And this privilege, so the argument goes, undermines any sort of anti-realist Pessimistic Induction from the history of science.

Ludwig Fahrbach (2011) has recently developed this line of argument in an attempt to blunt the threat posed by the version of the Pessimistic Induction I attributed to Putnam in the previous chapter. Fahrbach insists that this Pessimistic Induction is based on a miscalculation of the evidence from the history of science. Specifically, Fahrbach claims that those advancing the Pessimistic Induction fail to take account of the exponential growth of science. He argues that, given that most scientific research has been done in the last sixty years or so, the many past theories that have been discarded can hardly compare to the many theories developed in the last sixty years that are still accepted today. And these more recently developed theories are supported by far more data than were the theories they replaced. It is in this sense that today's scientists are epistemically privileged. Thus, according to Fahrbach, an induction from the history of science, that is, the *whole* history of science, *supports* realism. Fahrbach is thus led to draw an optimistic conclusion from the history of science. Incidentally, he is not alone in drawing such an inference. Robert Nola (2008) also thinks that the history of science supports an optimistic induction rather than a Pessimistic Induction.[1]

My aim in this chapter is to evaluate this new realist strategy for addressing the Pessimistic Induction, the appeal to epistemic privilege. I will focus primarily on Fahrbach's argument, but my criticisms are generalizable and apply to other similar strategies. I begin by explaining Fahrbach's argumentative strategy, the appeal to the exponential growth of science, and identifying what exactly it is intended to prove. Then I critically scrutinize the viability of the realists' appeal to the exponential growth of science. I aim to show that earlier generations of scientists could have constructed a similar argument, but one that aimed to show that the theories *they* accepted were likely true. The problem with this is that from our perspective on the history of science, we know that argument is flawed. Consequently, we should not be persuaded by Fahrbach's argument. It seems quite plausible that later generations of scientists will have a similar perspective on us to the one we have on our predecessors. I argue that Fahrbach fails to identify a difference that matters between today's theories and past theories. Realists, though, need to find such a difference if they are to undermine the Pessimistic Induction, at least if they are to maintain that the fate of past theories is irrelevant to an assessment of today's theories. Finally, I examine a similar argument against the Pessimistic Induction developed by Michael Devitt. Unlike Fahrbach, Devitt believes that developments in methodology undermine any inference from the history of science to the fate of today's best theories. He is thus suggesting that it is the better methods that make today's scientists epistemically privileged in a way that their predecessors were not. Devitt's argument, though, is prone to a criticism similar to the criticism I raise against Fahrbach.

---

[1] In his attack on the Pessimistic Induction, Nola appeals to the causal theory of reference, a theory that Hilary Putnam is partly responsible for developing. Michael Devitt (2011), though, rightly notes that it is problematic to attempt to address the Pessimistic Induction by appealing to a theory of reference, given the disagreement among philosophers about theories of reference (289). Devitt is a realist. His concern is that this appealing to a specific theory of reference does not provide the strongest defense of realism.

The Exponential Growth of Science

Fahrbach aims to show that we do not have strong *inductive* grounds for believing that our current best theories are probably false and thus apt to be discarded in the future. This is the form of the Pessimistic Induction that concerns him. It is an inductive argument, and though Fahrbach attributes the argument to Laudan, as we saw in the previous chapter, this form of the Pessimistic Induction was developed by Putnam. Despite the mistaken ascription, it is worth examining Fahrbach's concerns, for he draws attention to some of the key issues that are at stake in the debate about the relevance of the history of science to the realism/anti-realism debate.

As we saw in the previous chapter, realists have adopted a variety of strategies in their efforts to undermine this powerful anti-realist argument. Some suggest it is a fallacious argument, though there is some debate about what fallacy it commits. Magnus and Callender (2004), for example, claim that the Pessimistic Induction commits the base rate fallacy. Marc Lange (2002) suggests that the Pessimistic Induction commits the turnover fallacy. Peter Lewis (2001) claims that it commits a false positive fallacy. And others have attempted to cut down the long list of once successful but now discarded theories in an attempt to reduce the evidential base supporting the anti-realists' inference (see, for example, Hardin and Rosenberg 1982). In this way, realists aim to show that anti-realists exaggerate the ratio of once successful but now false theories to successful true theories (see Psillos 1996, S307). Others have attempted to distinguish between those parts of past successful theories that are responsible for their success and the extraneous parts in an effort to show that only the latter are discarded when one theory replaces another (see Kitcher 1993; Psillos 1999). This strategy is aimed at isolating the threat posed by the Pessimistic Induction.

Fahrbach has recently developed a new strategy for addressing the threat posed by the Pessimistic Induction. Drawing on the work of Derek de Solla Price (1963), Fahrbach notes that science has been growing exponentially, doubling in a variety of specific measurable ways every fifteen to twenty years. For example, the number of people working in science and the number of scientific articles published double every fifteen to twenty years. This means that contemporary science accounts for a disproportionately large portion of the total quantity of scientific research ever produced. Seen from this perspective, *the history of science from 1650 to 1950* is but a small part of science. Fahrbach provides us with a useful figure. He suggests that, given the exponential growth of science, 80 percent has been done in the last sixty years (see Fahrbach 2011, 139). The history of science before 1950 represents a mere 20 percent of the science ever produced.

These figures come from Price's (1963) work in scientometrics, a field concerned with quantitative studies of science. Price also notes that the number of scientific journals has been increasing exponentially (see Price 1963, 9, figure 1). In the early 1960s, Price predicted that the growth rate of science would slow down, as the then-current growth rate could not be sustained for much longer. He thought that we would reach a saturation point very soon. Fahrbach argues that the exponential growth rate has continued, contrary to Price's prediction. And Fahrbach attributes the continuing growth to the even more rapid growth rate in developing countries like China and India (see Fahrbach 2011, 148).

In an effort to show how the growth rate of science changes our perspective, Fahrbach considers the various mature theories of light that have been rejected. It is an impressive list of failures: Newton's particulate theory, Fresnel's wave theory, and Maxwell's theory. The latter was ultimately replaced by the photon theory, the theory that is currently accepted (Fahrbach 2011, 141–142). When we consider these theories without regard for the exponential growth of science, we may be dismayed and expect that sometime in the future we will likely discard the currently accepted photon theory. But Fahrbach insists that when we consider the exponential growth of science, we realize that these past successful but false theories were all developed and rejected in a time period during which less than 20 percent of all science was produced (compare figure 1 on page 142 with figure 3 on page 150 in Fahrbach 2011). As far as Fahrbach is concerned, it is irrational to draw an inference about the fate of our current best theories from such an idiosyncratic and unrepresentative sample. Indeed, Fahrbach is not the only one to raise the concern that the Pessimistic Induction is based on an unrepresentative sample. Moti Mizrahi (2013) also raises this concern, as we will see shortly.

Figure 3 provides a graphic illustration of Fahrbach's point about the exponential growth of science. The upper part of the figure shows the history of modern science from 1650 to 2010, with each sixty-year period represented by an equal-sized space. The lower part of the figure

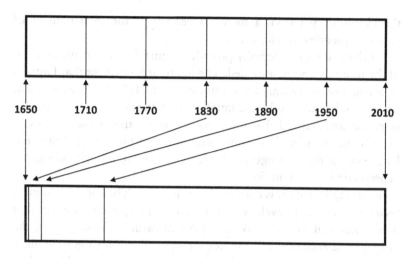

Figure 3 Two views of the history of science
Upper Figure: The history of science, 1650–2010, divided into sixty-year segments of equal size.
Lower Figure: The history of science, 1650–2010, divided into segments in proportion to the amount of research produced.

shows the same time period, but each sixty-year period is represented proportional to the amount of scientific research produced in that period, assuming science has been growing exponentially, as Fahrbach claims. Thus, the most recent sixty years covers 80 percent of the space in the lower part of the figure (see Figure 3).

The exponential growth rate of science has important implications for the Pessimistic Induction, implications that Fahrbach suggests undermine the argument. First, the exponential growth of science undermines the Pessimistic Induction because the inductive base from the part of the history of science from which anti-realists draw their sample is so small compared to the many theories scientists currently work with, many of which were developed in the last hundred years or so (see Fahrbach 2011, 149; see also Mizrahi 2013). So, even though scientists may have discarded 85 percent of the theories they developed in the past, say before 1950, most of the theories they have developed are still accepted today. That is, assuming that the development of theories is spread proportionally with the distribution of articles and scientific personnel, the 85 percent of discarded theories is drawn from a mere 20 percent of the total number of scientific theories developed.

Fahrbach does not actually provide a figure for how many theories of the past have been discarded. The figure of 85 percent that I appeal to is equal to approximately six out of seven. This figure, as noted in Chapter 5, has acquired some significance in the debate in the literature since Laudan claimed that for every successful theory the realist can identify, he can identify six once successful but now rejected theories. I will assume that this figure is more or less correct, though not much depends on this assumption.

Assuming Fahrbach is correct about the growth of science and most theories have been developed in the recent past, it seems likely that most successful theories have not been discarded. Consequently, the history of science does not provide strong evidence for the pessimistic conclusion that our current theories are also likely to be discarded. Hence, if Fahrbach is correct, Laudan could not come up with six once successful but now rejected theories for every successful theory that scientists still work with.

Let us consider what the evidence reveals, assuming Fahrbach is correct about the exponential growth of science. If 85 percent of the theories developed before 1950 have been rejected, and these constitute only 20 percent of the theories ever developed, then it seems that only 17 percent of the theories ever developed have been rejected. Perhaps we should add to this number some small number of theories that have been developed since 1950 that have already been rejected. Let us assume that about 4 percent of the theories developed since 1950 have been discarded. Given the exponential growth of science, the theories developed since 1950 constitute 80 percent of the theories ever developed. Hence, an additional 3.2 percent of all once successful theories have been rejected (80% × 4% = 3.2%). Thus, assuming these numbers are accurate, only about 20 percent of all theories have been rejected, a number that hardly supports a Pessimistic Induction (17% + 3.2% = 20.2%).

Second, Fahrbach appeals to the exponential growth of science in order to convince us that the theories we accept today are *fundamentally* different from past scientific theories in a way that warrants our being optimistic about the fate of our current best theories even though many successful theories were discarded in the past. An obvious way to undermine any ampliative inductive inference is to show that the past cases that form the inductive base differ fundamentally from the cases in the target group, whose properties are not directly known. This aspect of Fahrbach's defense of realism is not explicitly developed. But if realists are to blunt the threat of the Pessimistic Induction, they must identify some significant difference between today's theories and past theories. Without an argument to the effect that there is a *fundamental* difference between the theories we currently accept and the once successful theories we have since rejected, we have little reason to believe that today's theories will not end up on the pile of ruins to which Poincaré drew our attention.

Fahrbach, though, argues that the significant difference is that today's theories are supported by substantially more data than theories developed earlier in the history of science (see Fahrbach 2011, 149). If 80 percent of all scientific research has been done in the last sixty years, then the theories developed in the last sixty years that are still accepted today are supported by a substantially larger body of data than the once successful but now rejected theories developed two hundred or even one hundred years ago. There was just far less scientific research done then, and consequently less data supporting those theories. In Fahrbach's words, "more scientific work results in the discovery of more phenomena and observations, which, in turn, can be used for more varied and better empirical tests of theories" (Fahrbach 2011, 149). Thus, it seems that today's scientists really are epistemically privileged compared to their predecessors.

Indeed, Fahrbach suggests that if today's theories were not significantly superior to the theories we accepted in the past, then, given the exponential growth of scientific research, we should expect to see far more cases of revolutionary theory change (see Fahrbach 2011, 151). Consider, for example, the various theories of light. If there were four different theories of light accepted between 1600 and 1915, then, given the growth of scientific research since 1915, if our current theories are no better than the theories they replaced, we should have experienced more than four times the number of revolutions in that field since 1915. Since this has not happened, Fahrbach argues, we have compelling evidence that today's theories are superior to the theories of the past.

In summary, Fahrbach's strategy is to undermine the Pessimistic Induction in two ways. First, he insists that the bulk of evidence does not support an induction of the sort the anti-realist makes. The anti-realist was led to think otherwise because she considered an unrepresentative sample, a sample drawn from the early history of science, before many of our best theories were developed. Second, Fahrbach wants us to believe that there is some sort of significant difference between older scientific theories and today's theories, a difference that makes an inductive inference of the sort the anti-realist draws unwarranted. He claims that the key difference is that today's theories are supported by much more evidence than theories of the past. If the evidence gathered in support of more recently developed theories stands in a four-to-one ratio to the evidence gathered in support of theories developed more than sixty years ago, as Fahrbach suggests, there may be as much as four times the evidence supporting today's best theories compared with the evidence supporting older theories.

Assessing the Realists' Appeal to the Exponential Growth

It is worth drawing a distinction between two goals that Fahrbach has in appealing to the exponential growth of science. On the one hand, he aims to show that the Pessimistic Induction is a flawed argument and poses no real threat to realism. On the other hand, he aims to show that an induction from the history of science actually *supports* realism. That is, he believes that the history of science gives us good reason to be optimistic about our current best theories. The exponential growth of science, Fahrbach insists, is the key to understanding why history is on the side of the realist. In this section, my principal aim is to address Fahrbach's second goal, his attempt to draw support for realism from the history of science. I aim to show that the history of science does not support realism despite the fact that science has been growing exponentially.

Let us consider why the exponential growth of science does not offer support for scientific realism. The realist should realize that every generation of scientists (and philosophers of science) could run an argument similar in structure to the argument developed by Fahrbach. And this is a problem for Fahrbach's defense of realism. Consider the scientists and philosophers of science who lived and worked between 1890 and 1950. These scientists could also argue (then) that they were responsible for 80 percent of the scientific research ever produced. And they too would likely note that few of the theories developed during their era had been thrown out, at least by 1950. Indeed, today we see the science of their era differently, but they did not have the perspective that we have now.

And the generation before this one, the group of scientists who lived and worked between 1830 and 1890, could have constructed a similar argument. They would be impressed by the fact that they were responsible for 80 percent of the science ever produced, and they too would note how few of the theories developed during their lifetime were thrown out (then). But we know that despite what these scientists thought, many of the theories they developed have since been thrown out.[2]

If this line of reasoning is correct, then the exponential growth of science does not offer the realist grounds for drawing an optimistic induction about our current best theories from the history of science. Provided that we see ourselves as similar to the scientists of the generations that preceded us, we must take seriously the fact that once successful theories are often discarded later, on the grounds that they are false. Thus, the exponential growth of science does not offer support for scientific realism in the way Fahrbach suggests.

The Pessimistic Induction asks us to see ourselves as similar to the scientists of the past. We are not to be Whigs or deluded, assuming that we are not prone to make the same sorts of mistakes that they were prone to make. In short, we are not to *assume* some sort of epistemic privilege for today's scientists. Like their scientific predecessors, today's scientists have developed theories that are successful, just as the theories their predecessors developed *were* successful. The science of today enables us to see the many faults in the theories of earlier generations who were impressed by the success of the theories they accepted. In turn, contemporary scientists should expect that their scientific offspring will look back at their theories with the same attitude they have toward the theories of their predecessors. Their offspring will see that many of today's successful theories will have been discarded and replaced by new theories that today's scientists never even entertained accepting, theories that are currently unconceived. The reader should be reminded of Kyle Stanford's (2006) Argument from Unconceived Alternatives, discussed in the previous

---

[2] This line of argument cannot be run back indefinitely. At one point in the history of science, the pool of successful theories may be too small to warrant a conclusion about the prospects of future scientific theories. But let us set this concern aside.

chapter. The Pessimistic Induction is thus designed to aid us in recognizing the similarities between our predicament and the predicament of our predecessors.

The key to undermining the Pessimistic Induction is to block any sort of inference from past theories to present theories. This requires identifying some sort of feature common to today's theories that the successful but now discarded theories of the past lacked. That is, the key is to insist on some sort of epistemic privilege. Fahrbach fails to identify a fundamental difference between today's theories and the theories of the past, one that would undermine the ampliative inference made in the Pessimistic Induction.

What would such a difference have to look like?

We know that today's scientists are different from the generations that preceded them in a number of respects. Today's scientists have instruments and methods that were unthought of and perhaps even unthinkable by the generations before. As a result, scientists today are able to study whether the conditions on planets in distant solar systems can support life and to reconstruct the genomes of the long-extinct mammoth and Neandertal. And scientists have achieved degrees of accuracy in measurement and prediction that were unachievable before. The achievements of scientists are undeniably impressive.

But as we saw earlier, the problem is that previous generations could construct similar arguments with respect to the generations that preceded them. They had instruments and methods their predecessors could not fathom, and they achieved degrees of accuracy never achieved before. The pattern is clear. What looks innovative to our predecessors does not look innovative to us. And similarly, what looks innovative to us will not look so innovative to our offspring. Henri Poincaré made a similar observation: "Every age has scoffed at its predecessor, accusing it of having generalised too boldly or too naïvely. Descartes used to commiserate the Ionians (*sic*). Descartes in his turn makes us smile, and no doubt some day our children will laugh at us" (Poincaré 1905/2001, 109).

In order to develop an appreciation for the challenge realists face, it is worth considering two different strategies that the realist might appeal to in an effort to build a case for the claim that today's successful theories are different from past successful theories that we have since determined to be false. One strategy is for the realist to appeal to the increasing accuracy of theories. The second strategy, Fahrbach's strategy, is to appeal to the fact that today's theories are supported by far more data, and are thus far better corroborated. Both of these considerations might seem to suggest that today's scientists are in fact privileged, even if their predecessors were not. But let us consider each of these strategies in detail and see why they ultimately fail.

First, consider the increasing accuracy that has been achieved throughout the history of science. Each generation can claim that their theories are more accurate than their predecessors' theories. It is tempting to regard this feature of science as evidence that we have finally got things right. It looks like we are converging on the truth. But the problem is that the increasing accuracy we see throughout the history of science does not allow the realist to establish a clear break between the science of today and the science of the past. Even though today's theories are more accurate than any theories ever produced to date, there is no reason to think that an even greater degree of accuracy may not be achieved in the future. And appeals to the exponential growth of science give us no reason to think that the greater accuracy of the future will not be a consequence of our discarding today's theories and replacing them with hitherto unconceived theories. Moreover, if greater accuracy in the future is a consequence of today's best theories being *replaced* by new theories, theories that make radically different assumptions about the unobservable reality underlying the appearances, the realist gains nothing. What the realist needs is for the increasing accuracy to be accompanied by theoretical continuity. But Fahrbach has not provided an argument in support of this point.

Second, let us consider Fahrbach's appeal to the fact that today's theories, unlike the theories of the past, are supported by far more data, perhaps four times the data, given the exponential growth of science. Unfortunately, even this sort of argument could have been made by each preceding generation of scientists. Each generation of scientists can claim that their theories are supported by far more data than their predecessors' theories were. In fact, given the exponential growth of science, each generation will claim that their theories are supported by as much as four times more data than the previous generation's theories were. Hence, Fahrbach has still not distinguished today's theories from past theories.

The realist needs to find a difference that would distinguish today's theories from previous theories, but not the previous generation's theories from the theories of the generation that preceded it. This is no small feat, and certainly not a feat that Fahrbach has accomplished. The history of science is thus far more threatening to realism than Fahrbach realizes.

I suspect that there are two considerations misleading Fahrbach in his reasoning, considerations that may mistakenly be leading him to think that he is more warranted in being so optimistic about contemporary theories. I think we can easily be misled to believe that things are different now because many of the theories developed in the last sixty years continue to be accepted. They *seem* so resilient, and thus so different from past theories. But their resilience should not mislead us. It may take time for our current theories to be replaced by other theories. In fact, the history of science suggests that it does take some time.

Fahrbach also seems to be impressed by the great quantity of theories developed in the recent past that continue to be accepted today. But we must keep our eyes on what matters. It is not the number of theories that matter, at least not the number alone. What really matters is the endurance of theories. If we see our best theories lasting longer and longer, then we may have some grounds for thinking that scientists are getting better at developing theories. But Fahrbach does not provide evidence for this claim.

Devitt's Appeal to Developments in Methodology

Fahrbach is not the only realist who attempts to sever the link between the fate of theories developed in the early history of science and what we can expect of our contemporary theories. Devitt employs a similar strategy in his attack on the Pessimistic Induction and Stanford's new induction (see Devitt 2011). Roughly, Devitt argues that the best explanation for the success of our current theories, and the fact that they are superior to the theories they replaced, is that they were developed and tested with the aid of *better methods*, methods developed more recently in the history of science than those used to develop and test the many theories that were discarded earlier in the history of science. As far as Devitt is concerned, it is no surprise that earlier theories needed to be replaced. But our current theories are different, having been developed and tested with the aid of *superior* methods. Contemporary scientists are thus epistemically privileged by virtue of the superior methods they employ. And a Pessimistic Induction from the history of science is thus undermined. What has happened in the past is of little relevance to what is likely to happen to our present theories.

In this section, I aim to show that developments in methodology cannot support Devitt's claim about the unlikely occurrence of revolutionary changes of theory in the future. I argue that it is likely that scientists will continue to develop new methods in the future, and *some* of these methods will likely lead scientists to generate data that cannot be reconciled with the currently accepted theories. Consequently, contrary to what Devitt suggests, I argue that there is reason to believe that our current best theories may be replaced in the future by different theories that make radically different assumptions about unobservable entities.

Devitt believes that the anti-realists advancing the Pessimistic Induction seem to suppose "that we are *no better* at finding out about unobservables now than we were in the past" (Devitt 2011, 290; emphasis in original). But he claims that "just the opposite seems more plausible: we are now *much better* at finding out about unobservables" (see Devitt 2011, 290; emphasis in original). In fact, Devitt claims that "a naturalized epistemology would surely show that science has for two or three centuries been getting better and better at this" (Devitt 2011, 290). Specifically, Devitt claims that "scientific progress is, to a large degree, a matter of *improving scientific methodologies often based on new technologies that provide new instruments* for investigating the world" (Devitt 2011, 290; emphasis added). It is worth highlighting the broad conception of methodology that Devitt is working with. He connects developments in instrumentation with developments in methodology. But he does not think that developments in instrumentation exhaust the range of new methods in science. He also believes that there are methods of reasoning, for he makes a passing remark about the methods of "nondeductive ampliative inference" (see Devitt 2011, 287).

Given the developments that have been made in methodology, Devitt argues, "we should expect an examination of the historical details to show improvement over time in our success ratio for unobservables" (Devitt 2011, 290). That is, a careful look at the history of science should show that scientists are more likely to be right about the unobservables they posit now than their predecessors were three hundred years ago. Devitt believes that this argumentative strategy, this appeal to developments in methodology, undermines the anti-realists' inference in the Pessimistic Induction.[3] It severs the link between the once successful but now discarded theories of the past and today's successful theories, thus undermining the warrant for an inference from (i) a consideration of the shortcomings of past theories to (ii) expectations about the shortcomings of today's theories. According to Devitt, though past theories may have been prone to posit the existence of entities we have subsequently come to believe do not exist, given the developments that have been made in methodology, those past failures are irrelevant to evaluating the prospects of today's theories (see Devitt 2011, 290).

Devitt is surely correct to claim that many of the more recently developed methods in science are superior to the methods developed in the past. The evidence for their superiority is extensive. Most notably, more recently developed theories, developed with the aid of more recently developed methods, are generally more accurate than their predecessors. Further, some of the instruments developed in the recent history of science have extended the range of what scientists can observe quite considerably. Often, with the development of new methods, new phenomena are disclosed. For example, with the discovery of x-rays and the development of x-ray technology, crystallographers were able to study the molecular structure of crystals, and thus collect data that were hitherto unavailable to them (see Law 1976). And radio astronomy has greatly expanded the range of data astronomers can collect, opening up new horizons to study (see Edge and Mulkay 1976).

But it is a mistake to *presume* that the superiority of more recently developed methods increases our knowledge of *unobservables*. To assume so would be to assume the truth of realism rather than provide evidence in support of it. I grant that more recently developed methods are superior with respect to increasing our knowledge of the phenomena. Scientific theories, though, still posit unobservables. By the term "unobservable," I merely mean to capture the range of entities that Stanford describes as "too fast or too slow or too rare or take place on too grand a scale for us to engage with in ordinary ways" (Stanford 2006, 3). In order for Devitt's argument to offer support for the type of

---

[3] Devitt also believes that this is the best argumentative strategy against Kyle Stanford's Argument from Unconceived Alternatives.

realism he aims to defend, we need a reason to believe that developments in methodology in fact increase our knowledge of unobservables as well. Devitt's appeal to developments in methodology does not provide such a reason.

Further, the history of science is filled with many examples of developments in methodology, and many of these methods were inconceivable to earlier generations of scientists. New methods almost invariably introduce new data, and sometimes the new data cannot be reconciled with the accepted theories. Importantly, the new data are not just more of the same kind of data that scientists had been collecting in the past. Rather, the new data are sometimes qualitatively different from the data that scientists had considered to date. Galileo's telescopic discoveries are of this kind: the moons of Jupiter and the phases of Venus, for example. This was also the case with the discovery of x-rays (see Law 1976).

I suspect that many realists assume that there is a pattern in scientific development where many, if not most, of yesterday's unobservables have become today's observables. But this is not obviously the case. Interestingly, some of the new observables that new instruments disclosed were *not* regarded as unobservable entities before they were discovered. Rather, they were not even hypothesized to exist. That is the case, for example, with some of the other phenomena that Galileo discovered with the telescope. Jupiter's moons, for example, were not hypothesized to exist prior to their discovery. X-rays are similar in this respect. So, contrary to what some realists may want us to believe, we should not think of new methods and instruments that disclose new observables as rolling back the range of unobservables. Often what they do is extend the range of observables to hitherto completely unknown phenomena.

As we saw in Chapter 5, Nicholas Rescher provides a vivid way to understand this pattern in scientific discovery. Rescher (1987) suggests that scientists are constantly exploring "new regions of parametric space." With the aid of new instruments and methods, scientists are collecting data on variables about which they have not collected data before (see Rescher 1987, 10–14).[4] Sometimes

---

[4] As we saw in the previous chapter, Rescher (1987) is a realist, but he wants to separate the progress in science from convergentism. That is, though he thinks that science is not aptly described as converging on the truth, he does not think that this admission undermines the claim that science progresses.

the new data collected with the aid of a new method can be reconciled with the accepted theory. That is, sometimes the new data merely augment the existing picture of the world suggested by the long-accepted theory. Indeed, much of the recent DNA sequencing of various animal species has merely reinforced scientists' previous theoretical speculations on the relations between various species. That is, much of the data garnered from DNA sequencing supports long-accepted views about the evolutionary history of many species. Such changes are consonant with Devitt's realism. No radical changes are necessary to account for these new data. But at other times, new data are not so readily integrated into the world picture supplied by our accepted theory. A specific example of this is the relatively recent discovery that there are two distinct species of African elephants. Though this discovery could be accommodated by supplementing the existing scientific lexicon, creating a new branch in the taxonomic tree where one did not exist before, this research required a more radical change to the accepted taxonomy (see Rohland et al. 2010). It was discovered that mammoths were more closely related to Asian elephants than either were to the two species of African elephants. Prior to this discovery, it was generally believed that the various species of elephant were more closely related to each other than any of them were related to mammoths. Indeed, sometimes scientists will need to develop a new theory, one radically different from the accepted theory, in order to account for new data gathered with the aid of new methods or instruments. And in these cases, Devitt's defense of realism is threatened.

Further, nothing that Devitt says about developments in methodology suggests that today's scientists are fundamentally different from past scientists in a way that would block the Pessimistic Induction. His criticism of the Pessimistic Induction is thus vulnerable to the same sort of argument I presented in the previous section against Fahrbach.

My aim in this chapter has been to examine the promise of one of the most recent realist attempts to blunt the threat posed by a particular version of the Pessimistic Induction, the version that I attributed to Putnam in the previous chapter. This realist strategy involves either a tacit or explicit appeal to epistemic privilege. Fahrbach argues that most of the scientific research that has ever been done has been done in the last sixty years, and few theories developed in the last sixty years have been discarded. Consequently, he claims, there is little evidence supporting the Pessimistic Induction. In fact, Fahrbach thinks that the history of science actually supports an optimistic induction. Further, he claims that because of the exponential growth of science, today's theories are supported by much more data than the theories developed earlier in the history of science were. Today's scientists, he claims, are thus in a significantly different position than the position occupied by earlier scientists.

Though this line of reasoning has some *prima facie* plausibility of holding out hope for realism, on further reflection it is clear that earlier generations of scientists could have constructed a similar argument, given the exponential growth of science. But we know that many of those previous generations were clearly mistaken, for many of their once successful theories have since been discarded. As a result, we should expect that future scientists will look at us and our scientific theories with the same dismay with which we look at our predecessors and their theories. But however we might feel about our scientific predecessors, it is mixed with a deep appreciation for their accomplishments. Reflective scientists and philosophers of science are well aware that today's theories are as effective as they are only because they have been built on the ruins of yesterday's best theories.

Fahrbach has given us no compelling reason to think that there is some feature about today's best theories that separates them fundamentally from scientific theories of the 1700s and 1800s. Indeed, the anti-realist willingly grants that our more recently developed theories are better insofar as they enable us to make more accurate predictions and manipulate the world in hitherto unimaginable ways. But none of these accomplishments give us any strong reason to think that today's theories will not also be discarded in the future.

It is worth noting that revolutionary changes of theory in the history of science are not the only developments that threaten the realist's case. Throughout the history of science, there has also been a proliferation of new scientific specialties (see Rescher 1978, 229, table 3). I mentioned this in passing in the previous chapter. Importantly, the ontological assumptions of the theories used in one specialty are sometimes incompatible with those used in another scientific specialty (see Wray 2011, chapter 7). Given the inconsistency between theories used in different specialties, some of these theories are bound to be discarded in the future, for two inconsistent theories cannot both be true. So the challenges facing realists are greater than Fahrbach seems to realize.

7 | *The Nature of Radical Theory Change*

.

Part of the confusion in the contemporary realism/anti-realism debate is due to the loose way the term "theory change" is used. This ends up having important implications for how some philosophers have evaluated the threat posed by the Pessimistic Induction. Whether a Pessimistic Induction from the history of science seems well supported or not depends on which events one counts as instances of theory change.

In this chapter, I have two aims. First, I aim to provide a clear definition of the term "theory change," one that will clarify both what the Pessimistic Induction is intended to show and the sorts of changes in science that pose a threat to realism. The relevant contrast here is with a modification of a theory, where an accepted theory is merely extended. If all changes of theory are really just extensions of existing theories, then there is little basis for the anti-realists' skepticism about theoretical knowledge. But I aim to show that not all changes of theory are merely innocuous extensions of existing theories.

Second, I want to address a challenge raised by Ludwig Fahrbach. In a recent article aimed at undermining the Pessimistic Induction, Fahrbach argues that the Periodic Table is a hard case for the anti-realist because it has persisted more or less unchanged for so long. Fahrbach takes this as evidence that chemists have settled on the true theory in chemistry. I aim to show that Fahrbach is mistaken. There was a revolution in the early twentieth century in chemistry that affected the Periodic Table of Elements. This particular case is especially interesting, because, on the one hand, much was left unchanged after this change of theory. But on the other hand, as I will show, the changes that did occur were fundamental changes and profoundly affected scientists' understanding of the chemical world. Finally, because this change of theory affected chemistry in the last hundred years, it suggests that radical changes of theory are not a thing of past and do not only affect immature sciences. In this respect, realists cannot so easily dismiss this case as irrelevant.

What Counts as a Change of Theory

Larry Laudan's (1981, 33) famous list of discarded but once successful theories in "Confutation of Conyergent Realism" has often been taken to be the empirical evidence in support of a Pessimistic Induction (see, for example, Psillos 1999, 101–102 and 104–105). Though, as I argued above, the Pessimistic Induction is not part of Laudan's argumentative strategy against scientific realism (see Lyons 2002; Wray 2015a), numerous philosophers have read and continue to read Laudan's article as providing empirical support for a pessimistic conclusion about the likely fate of today's best theories (see, for example, Magnus and Callender 2004; Lipton 2004, 145; Chakravartty 2007; Mizrahi 2013, 3219).

Many realists who read Laudan as advancing a Pessimistic Induction have taken issue with his list in one way or another. Some critics have suggested that some of the examples on Laudan's list were not actually successful theories. Stathis Psillos, for example, has raised doubts about a number of the theories on Laudan's list, including the caloric theory of heat (Psillos 1999, 113). Psillos also suggests that "it is doubtful ... that the contact-action gravitational ether theories of LeSage and Hartley, the crystalline sphere theory and the theory of inertia enjoyed any genuine success" (Psillos 1999, 105). He also questions whether the humoral theory of medicine and the effluvial theory of static electricity were truly successful (see Psillos 1999, 108). In contrast, others argue that some of the theories on Laudan's list are approximately true, contrary to what Laudan suggests (see Psillos 1999, 103; also Hardin and Rosenberg 1982).[1] At issue is whether there is a basis for constructing a strong inductive argument in support of a pessimistic conclusion about the likely fate of today's best theories. It is now widely acknowledged that we need to go beyond Laudan's list. Much of the discussion, however, has been unclear and imprecise about what sorts of things count as radical changes of theory. I want to

---

[1] Philip Kitcher has an interesting response to the Pessimistic Induction along these lines. He claims that "we believe that Priestley was wrong, Lavoisier was wrong, Dalton was wrong, Avogadro was wrong, and so on. But we also think that Lavoisier improved on Priestley, Dalton on Lavoisier, Avogadro on Dalton. So while we do not endorse the claims of our predecessors we do support their sense of themselves as making progress" (Kitcher 1993, 137).

illustrate this by looking at two recent attempts to undermine the inductive form of the Pessimistic Induction advanced by Putnam.

Moti Mizrahi recently subjected the anti-realists' claim that most past successful theories have been discarded to an empirical test, surveying a sample of theories and scientific laws drawn from a variety of sources (see Mizrahi 2013, tables 1 and 2). In an effort to collect an unbiased sample to test the Pessimistic Induction, Mizrahi drew upon the online *Oxford Reference*, specifically the following sources: *A Dictionary of Biology*, *A Dictionary of Chemistry*, *A Dictionary of Physics*, and *The Oxford Companion to the History of Modern Science* (see Mizrahi 2013, 3220). Mizrahi even provided preliminary reports of his samplings of successful theories and laws from the history of science. The results are more optimistic than proponents of the Pessimistic Induction suggest (see Mizrahi 2013, 3222). This is preliminary work, so it would be premature for us to settle the issue on the basis of Mizrahi's sample of forty scientific theories and forty alleged scientific laws, but there is a concern with his test that is worth noting.[2]

On the one hand, Mizrahi is to be praised for his efforts to bring some rigor to the debate between realists and anti-realists where rigor has been lacking. But on the other hand, there is a problem with the way he proposes to test the anti-realist claim. Specifically, his appeal to scientific laws is irrelevant to an assessment of the viability of the Pessimistic Induction. This is so for two reasons. First, the discovery of many scientific laws does not lead to radical changes of theory. Boyle's Law, for example, merely identified a previously unnoticed relationship between two variables, the pressure of a gas and its volume (see Kuhn 1962/2012, 28). It is a normal scientific discovery. Hence, the discovery of laws, insofar as they relate to normal science, is irrelevant to assessing a Pessimistic Induction. Second, and more importantly, scientific laws often persist through theory change. Thomas Kuhn is quite adamant about the need to distinguish between theories and empirical laws for just this reason. According to Kuhn,

---

[2] Mizrahi has extended his attack on the Pessimistic Induction with additional empirical studies (see Mizrahi 2016). The additional data he presents, though, are difficult to evaluate, given the way he has reported them. For example, he provides a list of various "theoretical posits" that have been retained, but he does not always supply the corresponding list of theoretical posits that have since been rejected.

as science develops, [empirical laws] may be refined, but the original versions remain approximations to their successors ... Laws, in short, to the extent to which they are purely empirical, enter science as *net additions* to knowledge and are never thereafter entirely displaced. (Kuhn 1968/1977, 19; emphasis added)

Theories are a different matter. Kuhn claims that theories are, in certain essential respects, holistic. Consequently, he claims, "theories ... cannot be decomposed into constituent elements for purposes of direct comparison with nature or with each other" (Kuhn 1976/1977, 19; see also Duhem 1906/1954). Given the holistic nature of theories, we should expect changes of *theory* despite the fact that scientific laws can often persist through theory change. Thus, the fact that scientific *laws* tend not to be discarded or replaced even when there is a change of theory need not undermine the anti-realists' Pessimistic Induction. The Pessimistic Induction is concerned with *theories only*.[3]

Let us consider the second example. In a series of papers meant to blunt the threat posed by the Pessimistic Induction, Ludwig Fahrbach presents a great range of scientific discoveries that he claims continue to be accepted, contrary to the expectations of the Pessimistic Induction (see Fahrbach 2011; 2017). His list includes the following:

the Periodic Table of Elements,
the theory of evolution,
the conservation of mass-energy,
the germ theory of infectious diseases,
the kinetic gas theory,
"All organisms on Earth consist of cells,"
$E = mc^2$,
"Stars are gaseous spheres,"
"The oceans of the Earth have a large-scale system of rotating currents," [and]
"There was an ice age 20,000 years ago." (see Fahrbach 2007, 5050)

---

[3] It seems that the sorts of things that Structural Realists take as evidence for continuity, the mathematical formulas that express relations between types of phenomena, are like empirical laws of nature. That is, they are the sorts of things that we should expect to persist through theory change. And their persistence through changes of theory provide little evidence that our theories are true with respect to the claims they make about the underlying unobservables. This much the Structural Realist grants.

Contrary to what Fahrbach suggests, these are not all examples of theories. Rather, this is a disparate lot. Included here are hypotheses and claims, as well as theories. As a result, these are not all examples of the sorts of things an anti-realist would count as relevant to assessing a Pessimistic Induction. The Pessimistic Induction is concerned, after all, with *theories* only.

I believe that a more exact definition of the sorts of things that count as instances of theory change would enable both sides in the realism/anti-realism debate to better assess the threat posed by the Pessimistic Induction. Indeed, it would also aid us in clarifying what sorts of commitments anti-realists have.

In his efforts to undermine the Pessimistic Induction, Fahrbach poses a challenge to what he describes as "a truly ambitious anti-realist" (see Fahrbach 2017, 5060, Note 44). He asks the anti-realist to "conceive of an alternative to the Periodic Table of Elements that is entirely different from it, but also able to provide a systematic categorization, like the categorization by molecular structure" (Note 44). Fahrbach thus appeals to the Periodic Table as an instance of a theory that has persisted for a long time. Indeed, he takes it to be typical of contemporary theories. But unlike the various theories that appear on Laudan's famous list, which were developed when many of the sciences were not yet mature, the Periodic Table is a product of mature science. Consequently, he argues, we should not be surprised to find that it, and other more recently developed theories, are more resilient than the theories on Laudan's list.

I question whether the Periodic Table of Elements is aptly characterized as a theory (see also Scerri and Worrall 2001, 416 and 436). It may represent part of the content of a theory, but by itself it is not a theory.[4] Further, I think Fahrbach's challenge to provide a new systematization of the various chemical elements is a task for chemists, not philosophers of science. Nevertheless, as an ambitious anti-realist, I aim to show that his example of the Periodic Table of Elements does not support his criticism of the Pessimistic Induction. Not only is the

---

[4] Scerri claims that Mendeleev's Periodic Table of Elements is a "classification rather than a model or theory" (Scerri 2012, 276). Scerri and Worrall claim that "the periodic table is not itself a theory and therefore directly underwrites no prediction. It is the 'periodic law' lurking in the background, underpinning the table, that makes predictions if anything does" (Scerri and Worrall 2001, 416).

Periodic Table not a theory, I aim to show that its resilience hides a revolutionary change of theory in chemistry.

That a revolutionary change of theory should elude our attention should not surprise us, for, as Kuhn notes, revolutionary changes of theory are often rendered invisible (see Kuhn 1962/2012, chapter XI). When scientists rewrite textbooks to integrate the revolutionary changes of theory that do occur, they tend to emphasize the continuities through theory change. This is the way that scientific revolutions are rendered invisible. This is the Orwellian dimensions of theory change that Kuhn noted (see Kuhn 1962/2012, 166). Indeed, this tendency to emphasize the continuities is central to the realists' argumentative strategy (see, for example, Worrall 1989; Psillos 1999). In the next section, I articulate a conception of theory change that puts us in a better position to assess the threat that changes of theory pose to scientific realism.

Kuhnian Revolutions

Let me be clear about what I mean by theory change. Importantly, I want to get at a notion of theory change that would be relevant to a Pessimistic Induction. The Pessimistic Induction is an argument intended to show that most past successful theories have been shown to be false and consequently discarded, and that projecting into the future, we should expect that a similar fate awaits today's theories, even some of our best theories. So the notion of theory change relevant here must involve significant discontinuities between the discarded theory and its successor. Anything less than this would merely constitute a modification of an existing theory, and thus be irrelevant to the Pessimistic Induction. If most alleged instances of theory change turn out to be mere modifications, then it seems that realists may in fact provide a plausible account of the success of science.

In an effort to clarify what sorts of changes threaten realism, I will draw on Kuhn's mature account of scientific revolutions. Kuhn had originally described scientific revolutions as paradigm changes (see Kuhn 1962/2012). Roughly, he argued that a scientific revolution occurs when one paradigm or theory replaces another incommensurable paradigm or theory. It is the fact that the replaced paradigm and the new paradigm are incommensurable that makes paradigm changes revolutionary. The incommensurability is a consequence of two factors: (i) there are no overarching paradigm-independent standards by which to evaluate the competing theories and (ii) the terms and concepts used in the long-accepted paradigm and the new alternative paradigm do not cut nature at the same joints. One of Kuhn's favorite examples of a change in concepts accompanying a change in theory is the change from Newtonian mass to Einsteinian mass. Kuhn notes that "Newtonian mass is conserved; Einsteinian is convertible with energy. Only at low relative velocities may the two be measured in the same way" (see Kuhn 1962/2012, 102). Those who are resistant to the idea of revolutionary changes of theory tend to focus on the continuity between successive paradigms. Kuhn, though, believes that the continuity is often illusory, as this example illustrates. We should not be misled by the fact that the same *term* is used in both theories into thinking that there is extensive continuity between them. "Mass" does not mean the same thing. Because of such changes, the dispute between adherents of the long-accepted paradigm and of the new alternative paradigm can be complicated and contentious.

This characterization of revolutionary changes of theory, though, came under attack almost immediately after Kuhn published *Structure*. The process Kuhn described seemed to threaten the rationality of science (see Shapere 1964/1980; Scheffler 1967).[5] And the term "paradigm" was used in such an undisciplined way that it was never quite clear to Kuhn's critics what exactly he was claiming (see Masterman 1970; see also Conant in Cedarbaum 1983). As a consequence, Kuhn spent much of the rest of his career trying to clarify what he was trying to say in *Structure*. Despite the ambiguity inherent in his use of the term "paradigm," it did not stop others from using it.[6]

Ultimately, Kuhn settled on a different characterization of theory change (see Kuhn 1987/2000; 1991/2000; also Wray 2011). Instead of referring to theories as paradigms, he came to believe that theories are scientific lexicons. Each theory is a scientific vocabulary that orders the relevant concepts in specific ways, with very precise relationships between the concepts. Many of the central concepts of a theory are related to each other, as genus to species. For example, in the Ptolemaic

---

[5] Karl Popper even took issue with the rationality of normal science, as Kuhn characterized it (see Popper 1970). The dogmatic acceptance of a theory that Kuhn says characterizes normal science is anathema to Popper's critical rationalism.

[6] See Wray (2017) for an account of how social scientists responded to the term "paradigm."

theory in astronomy, some celestial bodies, but not others, are classified as planets. The class of planets includes the Moon, Mercury,
Venus, the Sun, Mars, Jupiter, and Saturn. Common to all these
celestial bodies is that they are wandering stars, having a motion
distinct from the motion of the fixed stars. In this respect, they form
a natural kind. Within the class of planets, further distinctions were
made between the superior planets and the inferior planets. The inferior planets were those situated between the Earth and the Sun; that is,
Venus and Mercury. The superior planets were those situated between
the Sun and the sphere of the fixed stars; that is, Mars, Jupiter, and
Saturn. The planets are contrasted with an alternative class of celestial
bodies, the fixed stars, which were alleged to move together on the
outermost sphere of the cosmos, circling the Earth in twenty-four
hours. The fixed stars thus form another natural kind. They are a
distinct species of celestial objects. An alternative theory, according
to Kuhn, would either (i) invoke different concepts, or (ii) define the
concepts differently than the long-accepted theory does, or (iii) relate
the concepts to each other in different ways than they are in the long-
accepted theory.

In his later writings, Kuhn described revolutionary theory changes as
involving lexical changes of a very particular sort.[7] Specifically, they are
those changes of theory that reorder scientific concepts such that the
relations between concepts in the long-accepted theory are no longer
preserved. In fact, a revolutionary change of theory involves the violation of what Kuhn calls the no-overlap principle (Kuhn 1991/2000, 94).
Essentially, what this involves is the introduction of changes to the
scientific lexicon such that former relations of genus and species are no
longer preserved. Instead, classes of objects that were previously regarded
as related to each other as species to genus are no longer regarded as
related in this way. For example, with the Copernican Revolution in
astronomy, the Sun was no longer regarded as a planet. Instead, it came
to be regarded as a star, and the distinction between fixed stars and
wandering stars was abandoned. The Copernican Theory introduced
other lexical changes as well. The definition of planet, for example,
changed from wandering star to satellite of the Sun.

---

[7] Kuhn also uses the term "taxonomic change" when he discusses theory change in
his later work. He thought of the network of concepts associated with or
constitutive of a theory as forming a taxonomy of the objects in the scientific field
the theory serves (see Kuhn 1991/2000).

Kuhn contrasts these sorts of changes to a scientific lexicon with changes that merely enlarge a scientific lexicon. For example, when scientists discover a hitherto undiscovered animal species, often they are able to accommodate the newly found species by merely extending the existing scientific lexicon, still preserving the structure of the accepted lexicon. For example, when a new species of frog is discovered, the class of amphibians remains intact. Even the various species of frogs may retain their relative relationships to each other in a taxonomy. All that is required to accommodate the discovery is to add another branch to the prevailing taxonomic tree. Such a discovery can be exciting, but it would not be a revolutionary discovery in the Kuhnian sense.

Kuhn regarded this new characterization of theory change as merely a clarification of his original account, presented in *Structure*. Even he came to realize that the notion of a paradigm was far from clear (see Kuhn 2000, 298). In his later writings, the term "paradigm" was reserved for those specific scientific accomplishments that become templates for solving other related outstanding scientific problems (see Wray 2011; see Kuhn 1977b, xix–xx). For example, drawing on Tycho Brahe's vast store of observational data, Johannes Kepler developed a model for the orbit of Mars. This model embodied Kepler's famous first two laws of planetary motion: the orbit of a planet is elliptical in shape with the Sun occupying one focus, and the planet sweeps out equal areas in equal times as it orbits the Sun. This planetary model became a template for solving related research problems. For example, it could be, and was, used to develop models for the orbits of other planets. In time, the paradigm aided astronomers in developing models for the orbits of satellites, that is, moons of planets, and for the paths of comets. This is how paradigms function in Kuhn's later writings.

Much of the apparatus associated with the earlier paradigm-related notion of revolutionary theory change was retained in Kuhn's more recently developed lexical change model of theory change. For example, he continued to believe that anomalies played a crucial role in the process of scientific change that ultimately lead to revolutionary changes of theory. He also continued to believe that scientific fields were often led into a state of crisis by anomalies that persistently resisted resolution or normalization. The appeal to lexical changes is meant to bring into focus the essential characteristic of radical theory change, the sort of change we associate with scientific revolutions. Normal science, on the other hand, can be conducted effectively with the conceptual resources supplied by the prevailing scientific lexicon.

The importance of such changes in scientific lexicons is that they lead to significant changes in our understanding of the world. Such changes, Kuhn felt, undermine the popular understanding of scientific progress that assumes that the growth of scientific knowledge is cumulative, with no setbacks. These sorts of radical changes to a scientific lexicon undermine the continuity assumed by the common view of scientific progress (see Kuhn 1962/2012, 3 and 7). Even if each change of theory increases our understanding of the phenomena, something Kuhn admits, such changes are not compatible with the claim that we are marching ever closer toward the truth about unobservable entities. More precisely, Kuhn believes that it is hard to ground the claim that, over the course of numerous changes of theory in a field, scientists have been getting increasingly closer to the truth in their description of the unobservables underlying the observables (Kuhn 1962/2012, 96–97).

Part of the reason that continuity is disrupted when there is a radical change of theory is the holistic nature of scientific lexicons. For example, during the Copernican Revolution, when astronomers changed the extension of the term "planet," certain generalizations that had been accepted were no longer believed to be true. Most obviously, the generalization that all planets orbit the Earth was no longer accepted. Given the new lexicon, no planet orbits the Earth. And the generalization that the superior planets are situated between the Sun and the fixed stars was rendered problematic. Given the new lexicon, the superior planets are those that lie between the Earth and the fixed stars. Every radical change of theory has similar effects on what scientists believe.

The State of Chemistry at the Dawn of the Twentieth Century

I now want to examine a revolution in chemistry in an attempt to address Fahrbach's challenge. The revolution in chemistry that concerns me is not the one that has gained the most attention from philosophers of science, the revolution ushered in by Antoine Lavoisier that led to the replacement of the long-held phlogiston theory by the oxygen theory of combustion. This episode has been discussed extensively by philosophers of science (see, for example, Kuhn 1962/2012, 53–57 and 70–72; Thagard 1990; Pyle 2000; Hoyningen-Huene 2008; Kusch 2015). It is often referred to as *the* Chemical Revolution.

The revolution that concerns me is the discovery of the relevance of atomic number to the ordering of the elements in the Periodic Table. Part of my aim is to undermine one of Fahrbach's key examples of a long-stable theory that he takes to support scientific realism. The stability in this case is an illusion, created in part by the persistence of the Periodic Table of Elements through a significant change of theory. In addition, though, I want to draw attention to an example of a change of theory of the sort that is relevant to assessing the Pessimistic Induction.

Eric Scerri provides some useful background on this episode in the history of science (see Scerri 2007; 2011; 2013). He draws attention to the important ways in which this change had profound implications for both chemists' understanding of the elements and the structure of the Periodic Table of Elements. I will rely heavily on his work to support my claim about an overlooked revolution in early twentieth-century chemistry. My contribution will be in highlighting the way in which this episode is a radical change of theory in Kuhn's sense. This will also address Fahrbach's challenge to the ambitious anti-realist.

The nineteenth century was a golden age in chemistry. The century began with important research by John Dalton, which led to the discovery of the law of fixed proportions (see Kuhn 1962/2012, 78–79; 129–134). This provided an impetus for much research in the early 1800s. As Brett Thornton notes, "over fifty elements were discovered in the 19th century" (Thornton 2010, 86). And at a typical German university, chemistry grew from a discipline that was serviced by one professor in the 1820s to a discipline that required four professors by the 1890s (Ben David 1971, 125–126).

More important for our concerns is the Congress in 1860 in Karlsruhe, Germany. Over fifty chemists from across Europe gathered in Karlsruhe to settle various theoretical issues, in a deliberate attempt to move the field ahead (see Everts 2010). Specifically, they sought "more precise definitions of the concepts of atom, molecule, equivalent, atomicity, alkalinity, etc." (Weltzien 1860, cited in Hudson 1992, 123). Though the conference was a great success by many measures, a consensus on many issues was not reached. Some of those who attended adamantly insisted that "votes must not be taken on scientific questions" (see Ihde 1961, 85). At the end of the conference, the Italian chemist Stanislao Cannizzaro proposed that atomic weights be used as a means to classify chemical elements (see Kaji 2002, 5). Many chemists agreed, including Adolph Strecker and August Kekulé (see Ihde 1961, 85).

Though the issue of how to classify elements was more or less settled after the Congress in Karlsruhe, the organization of the various chemical elements was far from complete. Determining the atomic weight of a sample of a chemical element involved challenging work in the laboratory. And there was, in fact, some divergence of opinion on these empirical matters. J. W. van Spronsen provides a useful table for comparing the atomic weights assigned to a variety of elements by prominent chemists in the 1860s and early 1870s. Vanadium, for example, was assigned atomic weights ranging from 51 to 138 during this period. This illustrates the challenges that chemists still faced (see van Spronsen 1969, 57, table 2). Though chemists were working in what Kuhn would call a normal scientific tradition, the research problems they faced required ingenuity, patience, and often the development of specialized equipment and techniques.

But guided by the belief that atomic weight is the characteristic feature of each chemical element, a number of chemists embarked upon the task of developing ways to systematically organize the various elements. Scerri provides a useful summary of the various attempts to order the elements using atomic weight as the guiding principle (see Scerri 2011, chapter 4). And van Spronsen provides diagrams of a variety of periodic tables and other ways of ordering the elements developed in and after the 1860s, following the Karlsruhe Conference (van Spronsen 1969, chapter 5). The culmination of this research program was Dimitri Mendeleev's Periodic Table of Elements. In fact, Mendeleev developed numerous periodic tables in the 1800s, but he was committed to the idea that the defining feature of an element is its atomic weight (see Kaji 2002, 10). Based on his research and the first Periodic Table of Elements that he published in 1869, Mendeleev made a number of important observations about the chemical elements (see Hudson 1992, 130–131). They include: (i) "the elements, if arranged according to their atomic weights show a clear periodicity of properties"; (ii) "the arrangement of the elements ... in order of atomic weights, corresponds with their valencies"; and (iii) "the magnitude of the atomic weight determines the character of an element" (Mendeleev in Hudson 1992, 131). The Periodic Table thus not only organized the elements, it did so in a manner that revealed hitherto unknown features of the structure of the chemical world.

Using his various periodic tables, Mendeleev predicted a variety of hitherto unknown elements, and even predicted various properties of the elements based on the spaces left empty in the tables and their relative location with respect to other known elements (see Scerri 2011, 63–68). But as Scerri and Worrall have shown, the success of Mendeleev in this endeavor has been exaggerated (see Scerri and Worrall 2001). A number of his predictions proved misguided. In fact, Mendeleev's predictions of new elements seemed to be correct only half the time (see table 21 in Scerri 2011, 68). But clearly the method he used was useful in some respects, contributing to the discovery of some hitherto unknown chemical elements.

Not surprisingly, late nineteenth-century chemists faced some challenging anomalies. Perhaps one of the most famous is the debate about the relative placement of iodine and tellurium. Tellurium has an atomic weight higher than iodine's, yet given its other chemical properties, some chemists, including Mendeleev, thought that it should precede iodine on the Periodic Table of Elements. Scerri notes that "this step had already been taken by Odling and Lothar Mayer" (see Scerri 2007, 109 and 130; see also van Spronsen 1969, 113). Strictly speaking, this was a violation of the key principle of ordering the elements. But such practices are common in all sciences. Anomalous phenomena have to be dealt with in some way, and Mendeleev and other chemists felt that such a solution was reasonable, given their knowledge of the properties of the various elements. In fact, Mendeleev justified his decision on the grounds that the atomic weight assigned to one or both elements may not be correct (see Scerri 2007, 126). Other pairs of elements had posed similar problems for chemists, as long as they assumed that atomic weight was the key to classifying chemical elements, including, for example, potassium and argon, and cobalt and nickel (see Hudson 1992, 136; also 175; Heilbron 2005, 230).

Another anomaly or set of anomalies began to surface in the beginning of the twentieth century. Chemists were finding that some samples of elements that had the same chemical properties, and that were chemically inseparable, had different atomic weights. For example, "in 1906 Bertram Borden Boltwood ... was unable to separate ionium ... from thorium" (Hudson 1992, 170). If the principle of atomic weight were taken as the defining characteristic of an element, then these samples should have been regarded as distinct elements.

But the fact that they could not be chemically separated was uncharacteristic for distinct elements.

These anomalies in themselves did not necessarily signal a pending revolution in chemistry. After all, as Kuhn notes, every theory faces anomalies (see Kuhn 1962/2012, 145–146). Anomalies provide the research topics in a normal scientific tradition. And scientists sometimes choose to set some anomalies aside, to await the efforts of future scientists who may be better equipped, conceptually and technologically, to tackle the problems. Ultimately, though, these specific anomalies did contribute to bringing about a radical change of theory in chemistry.

The Early Twentieth-Century Revolution in Chemistry

The revolution in chemistry in the early twentieth century was not the result of scientists consciously seeking to radically change their field. Rather, it was a consequence of a series of research projects developed somewhat in isolation leading to a significant change in chemists' understanding of the world. The result, though, was a radical change of theory of just the sort that Kuhn would regard as revolutionary.

Perhaps most significant in this process was the discovery of atomic number, which is described in detail by Scerri (2007; 2011). This discovery follows the pattern that Kuhn identifies in *Structure of Scientific Revolutions*, in his analysis of the discovery of oxygen (see Kuhn 1962/2012, 53–57). Consequently, it is fruitless and futile to attempt to pinpoint who discovered atomic number and when exactly the discovery was made. Instead, we have to satisfy ourselves with identifying a range of dates between which it occurred. The various lines of research that led to the discovery were also quite different. Henry Moseley, for example, was using the then still new x-ray technology to analyze the structure of various elements (see Hudson 1992, 173; see also Alvarez et al. 2008, 92). Laboratory work by Moseley, Antonius van den Broek, and others ultimately led chemists to realize that they could order the elements according to their atomic numbers, with each element separated from the preceding element by one unit. This discovery was quite profound. It required some significant reconceptualizations, including the realization that the atomic weight of an element is determined not only by the number of protons in an atom, but also by the number of neutrons. The atomic number, on the other hand, is determined exclusively by the number of protons in an atom of the element. Most elements consist of several isotopes, which have the same number of protons but differ in the number of neutrons.

Interestingly, this change in the conception of chemical elements did not require a significant change in the order of most of the elements on the Periodic Table. As a consequence, one might mistakenly think that there was no revolution in chemistry in the early twentieth century. Indeed, in general, the continuities through instances of theory change can have this effect, masking over the revolutionary dimensions of the changes that occur when there is a change of theory in science.[8] This, I believe, is Fahrbach's mistake. Indeed, it is a common mistake made by realists.

Contemporaneous with this research on atomic number was another research program examining the various anomalous chemical elements that share the same chemical properties but differ with respect to atomic weight. Herbert McCoy and Frederick Soddy recognized that some of these alleged elements could not be chemically separated from other well-known elements (see Scerri 2013, 48). Soddy realized that these were not distinct elements, but rather variants of already known elements. He called the variants of an element "isotopes." They are variants that differ only with respect to their atomic weight and "the relatively few physical properties which depend upon atomic mass directly" (Soddy 1913, 400). They thus differ with respect to the number of neutrons in the atoms.[9]

In some respects, the concept of isotopes was a conceptual impossibility as long as chemists assumed that atomic weight defined chemical elements. After all, if two samples had different atomic weights, they

---

[8] This problem is exacerbated in chemistry, because some regard the Periodic Table of Elements as a theory. For example, Restrepo and Pachón argue that the various periodic tables are "just different representations of the same phenomena, different shadows of the same object – the Periodic Law" (Restrepo and Pachón 2007, 190). Insofar as the table did not change, or changed in only minor ways, one may be tempted to infer that there was no scientific revolution. This, though, misses an important point of Kuhn's account of theory change.

[9] This knowledge has been put to work in other scientific fields. For example, space scientists rely on our knowledge of isotopes to determine whether particular meteorites on Earth originated from Mars (see Treiman et al. 2000). And archaeologists have relied on our knowledge of isotopes in order to aid in the identification of skeletal remains, knowing that different cultural groups are exposed to different isotopes of lead, for example (see Carlson 1996).

were, by definition, different elements. So it is not surprising that it took some time, and some confidence, to assert the existence of such things as isotopes. The introduction of the concept "isotope" provided chemists with a conceptual tool to make sense of a perplexing phenomenon, specifically, the fact that samples of the same element can have different atomic weights. Isotopes are a classic example of something that does not fit the accepted lexicon, as long as scientists assume that elements are distinguishable by their atomic weight.

These two discoveries complemented each other. Once chemical elements were thought of as essentially defined by their atomic number, the notion of isotopes was no longer a conceptual impossibility. And once isotopes were recognized, many alleged elements were recognized to be merely variants of known elements, thus cleaning up the seemingly ever-expanding Periodic Table of Elements. For example, chemists came to recognize that what they had called "ionium" was actually just an isotope of thorium, not a distinct element.

The discovery of atomic number as the proper principle for ordering the elements also resolved the anomalous reversals that some chemists had made, like the reversal of tellurium and iodine, discussed earlier. Once chemists were committed to using atomic number as the ordering principle, the switch in order that they had made between tellurium and iodine became warranted. The new ordering, based on atomic number, ensured that "tellurium and iodine fall into their appropriate groups in terms of chemical behavior" (Scerri 2011, 83). So another set of anomalies was thus resolved with the discovery of atomic number.

The discovery also led to other unforeseen insights into the chemical world. Once the then-known elements were laid out in order according to atomic number, it became evident how many unknown elements there were, at least between hydrogen and uranium. In fact, once this was made clear, the race to find the seven missing elements became more directed. As Scerri explains,

while chemists had been using atomic weights to order the elements there had been a great deal of uncertainty about just how many elements remained to be discovered. This was due to the irregular gaps that occurred between the values of atomic weights of successive elements in the periodic table. This complication disappeared when the switch was made to using atomic number. Now the gaps between successive elements became perfectly regular, namely one unit of atomic number. (Scerri 2011, 80)

All seven of the missing elements between hydrogen and uranium were identified in the period between 1917 and 1945 (see Scerri 2007, 7 and 173–174; Scerri 2011, 80; Scerri 2013, xvi). This is a clear example of what Kuhn regards as normal science, filling in the details of a theory (see Kuhn 1962/2012, chapter III). The discovery of these elements did involve some challenging work. Thornton provides a list of the scientists who claimed to have discovered one of the missing elements, element 85, astatine, or "eka-iodine in Mendeleev's terminology" (Thornton 2010, 86). Still, the rate at which the seven elements were discovered is quite striking, and the efficiency with which they were identified is a consequence of taking the conceptual categories of the newly accepted theory as given.

Some revolutionary changes of theory are quite protracted, but a consensus was reached rather quickly in this case. As Robin Hendry notes, by "1923 the International Committee on Chemical Elements, appointed by the International Union of Pure and Applied Chemistry (IUPAC), enshrined nuclear charge as the determinant of the identity of the chemical elements" (Hendry 2012, 58). Perhaps even more telling is that by 1918, a textbook was published that included a periodic table that ordered the elements by atomic number (see Alvarez 2008, 92). This new chemical theory was now becoming part of the foundation of chemical education.[10]

In summary, the field of chemistry underwent a significant change of theory in the early twentieth century, when atomic number replaced atomic weight as the principle for ordering and identifying the chemical elements. This was a profound discovery. We risk misunderstanding the development of our knowledge in chemistry if we fail to see the episode for what it is.

In fact, it is a classic case of a Kuhnian revolution. In the process of addressing anomalies, chemists who were trained to see elements as defined by their atomic weight discovered that their theoretical assumptions were impediments to understanding the chemical world. The only way to normalize the anomalies was to introduce new concepts and a new conceptual understanding of what it is to be an element. As these changes were made in a piecemeal way, a new

---

[10] Textbooks play a crucial role in Kuhn's analyses of science and scientific change. They are the means by which scientists-in-training learn the scientific lexicon in their field (see Kuhn 1962/2012, 80–81; 164–165).

scientific lexicon emerged, one that took atomic number to be the defining feature of a chemical element.[11] It is worth emphasizing that this is quite a significant scientific revolution by Kuhn's standards. Kuhn makes clear that, though most scientists and laypeople are aware of the wide-ranging and protracted scientific revolutions, like the Copernican Revolution in astronomy and the Darwinian Revolution in biology, most scientific revolutions affect only a small group of specialists. This revolution in chemistry in the early twentieth century clearly affects the community of chemists as a whole. Granted, the Periodic Table of Elements retained much the same structure before and after the change of theory. But we should not be misled by this continuity and misunderstand the significance of the change in our chemical understanding.

It is worth emphasizing that, like the revolution in early modern astronomy, this revolution also involved a change in the extension and intension of key concepts. What counted as a chemical element after the revolution was determined by atomic number, not atomic weight or any other chemical or physical property. And given the change in intension of the concept "chemical element," it was determined that some suspected elements were not, in fact, elements after all. Rather, the various isotopes were deemed to be variants of already understood elements, not new hitherto undiscovered elements.

Also, some generalizations that were accepted before the change in lexicon were determined to be false after the revolutionary change of theory. For example, the generalization that *any two chemical samples differing with respect to atomic weight are distinct elements* was no longer accepted. Given the new lexicon based on ordering the elements according to their atomic number, some such samples were now regarded as isotopes of the same element.

I want to briefly consider an objection that I anticipate to the line of argument developed here with respect to the revolution in chemistry that I have identified. I suspect that some realists will focus on the continuities between the Periodic Table organized by atomic weight and the Periodic Table organized by atomic number. After all, the same elements appear on both tables, and the vast majority of the

---

[11] Scerri (2016) emphasizes the piecemeal way in which science advances. Indeed, this was part of his rationale for criticizing Kuhn's theory of scientific change. Scerri sees the process as more evolutionary than revolutionary. Elsewhere I have argued that Kuhn believed it could be both (see Wray 2011).

elements remain in the same relative position to the other elements. The realist is apt to argue that this is merely a case of modifying an existing theory.

Such a response fails to account for the nature of the changes that did occur. After all, as long as chemists were assuming that atomic weight was the defining feature of chemical elements, many more elements were thought to exist, and these were interspersed, or assumed to be interspersed, among the elements that did persist through the change of theory. And the discovery of isotopes, as already mentioned, was in some important sense impossible given the theoretical framework supplied by the Periodic Table that ordered the elements according to atomic weight. For these reasons alone, this case hardly resembles the case of the discovery of a new species of frog, discussed earlier.

Changes in experimental practices and the significance assigned to previous practices provide additional support for my claims about the magnitude and significance of this change of theory. For example, the laboratory techniques used to determine an element's atomic number differ from those used to determine its atomic weight. And whereas before the change of theory, a chemist might have described the laboratory manipulations to determine the atomic weight of a sample as a method or procedure for determining what chemical element the sample was, after the change of theory, the same laboratory manipulation could no longer be characterized in that way. This provides some substance to Kuhn's claim that after a scientific revolution, scientists work in a new world.

Part of my point in drawing on this example is to illustrate yet another Kuhnian revolutionary change of theory, the sort of change that is hard to reconcile with many forms of scientific realism. These are the sorts of changes relevant to assessing the Pessimistic Induction. Yet another part of my point is to address Fahrbach's challenge. Fahrbach wants us to believe that there is a great deal of stability and continuity in contemporary scientific theories. The Periodic Table of Elements was meant to be the hard case for the anti-realist. But I have argued that even that has been affected by a revolutionary change of theory. There is not, strictly speaking, a single Periodic Table of Elements that chemists, with the aid of physicists, have refined over time since the 1860s. Instead, there are at least two distinct Periodic Tables of Elements. The table organized on the principle of atomic numbers is not just a refinement of the table organized on the basis of atomic weight. To treat matters as if the table organized according to atomic number was just a refinement of the table organized according to atomic weight is to engage in Whig history of science and to be insensitive to the radical nature of the discovery and subsequent fallout of the changes that happened in chemistry in the early 1900s, when atomic number became the principle for ordering and identifying chemical elements.

We are now in a better position to understand what sorts of changes in science are relevant to an assessment of the Pessimistic Induction from the history of science. Whenever a scientific field endures a change of theory like the one described above, some degree of skepticism about theoretical knowledge is warranted. We are also in a better position to understand the limitations of the Pessimistic Induction, that is, the types of realism not threatened by radical theory change. Insofar as the anti-realist appealing to the Pessimistic Induction draws a distinction between empirical laws, which may persists through changes of theory, and theories, the Pessimistic Induction will not pose a real threat to a position like Structural Realism. Structural Realists, after all, grant that there are, and will likely continue to be, significant changes in the ontologies posited by our scientific theories. The sort of continuity that they are looking for involves the continuity of mathematical equations. But these equations are like empirical laws of nature, insofar as they merely identify regularities in the phenomena.

8 | Do *the Theoretical Values* Really *Support Scientific Realism?*

There is one final issue that deserves our attention in this critical assessment of realism. It is an evaluation of the realists' appeals to the theoretical values. The theoretical values – predictive accuracy, simplicity, and such – have played an important role in the realism/anti-realism debate in the philosophy of science.[1] Many realists argue or just assume that they are reliable indicators that a theory is likely true or approximately true with respect to what it says about unobservable entities and processes. Anti-realists disagree, claiming that these values are not *reliable* indicators of theoretical truth.[2]

In this chapter, I argue that these values are not capable of supporting the sorts of claims that realists lead us to believe they can support. Scientists are not warranted in inferring that the theories that embody these values are *likely true* or *approximately true*. Further, there is reason to doubt that these values are systematically connected to theoretical truth in the way that many realists suggest. I also argue that the theoretical values fail to provide scientists with the sort of practical guidance that some realists seem to suggest they can offer in choosing which theory to work with. Specifically, I argue that in the pursuit of better theories, it may not be the most expedient strategy to work with the theory that embodies the theoretical values to the highest degree.

I begin with a discussion of the theoretical values and the connection that many realists allege they have with theoretical truth. Then I review Larry Laudan's attack on realist appeals to the theoretical values, with the intention of bringing into focus what the realist must prove to support the sorts of inferences she seeks to make. At the same time,

---

[1] The theoretical values are also referred to as theoretical virtues or criteria of theory choice. I use these terms interchangeably.

[2] I will use the term "theoretical truth" to denote the fact that a theory is true or approximately true with respect to what it says about unobservables.

I aim to clarify the nature of Laudan's argument in "Confutation of Convergent Realism," as it has been so frequently misunderstood. I then argue that the theoretical values yield only ordinal rankings of competing theories, not the sorts of rankings that can support the sorts of inferences that realists typically want to make from them. Then I elaborate on the problems that realists face when they appeal to the theoretical values to ground their claims about our current best theories.

The Theoretical Values

The theoretical values are the values that scientists consider when they evaluate competing theories. Thomas Kuhn (1977) provides an often-cited list of these values. Kuhn claims that scientists appeal to the following values when assessing competing theories: simplicity, breadth of scope, fruitfulness, consistency, and predictive accuracy.[3] Other things being equal, scientists prefer theories that are simple, broad in scope, fruitful for further research, consistent with other theories one accepts, and capable of generating accurate predictions. Kuhn insists that these values have persisted throughout the history of science and have been appealed to in many scientific fields.

Realists and anti-realists alike recognize that these values are distinct from theoretical truth. Hence, in principle, a theory could embody these values and not be true or even approximately true with respect to what they say about unobservable entities and processes. But when faced with a choice between competing theories, scientists often do rely on these values. They do so because they are in no position to determine the truth value of the claims their theories make about unobservables directly. In contrast, scientists can directly ascertain whether or not a theory embodies these values.

---

[3] Helen Longino (1995) proposes an alternative list of values that overlaps somewhat with Kuhn's list. Her list includes empirical adequacy, novelty, ontological heterogeneity, complexity of relationship, applicability to current human needs, and diffusion of power (see Longino 1995, § 2). Longino argues that "in certain theoretical contexts, the only reasons for preferring a traditional or an alternative virtue are socio-political," which, she claims, shows that the traditional virtues cannot be "purely cognitive" (1995, 383).

Ernan McMullin provides a useful taxonomy of the theoretical values that offers insight into how the realist understands the relationship between the various theoretical values and theoretical truth.[4] He distinguishes between three types of theoretical values. First, McMullin notes that "some of the [theoretical] values ... function as goals of the scientific enterprise itself: predictive accuracy (empirical adequacy) and explanatory power are the most obvious candidates" (1993, 67). According to McMullin, each of these goals is "valuable in its own right, [and as] an end in itself" (67).[5] These values are constitutive of science.[6]

Second, McMullin notes that "other epistemic values serve as *means* to these ends; they help to identify theories more likely to predict well or to explain" (1993, 68). McMullin identifies "logical consistency ... and compatibility with other accepted knowledge claims" as examples of this type of theoretical value (68). These correspond to what Kuhn calls internal consistency and external consistency, respectively (see Kuhn 1977, 321–322; see also Kuhn 1962/2012, 184). Unlike the constitutive values, the values in this second set "are ... not goals in

---

[4] McMullin provides an alternative taxonomy of the "virtues of a good theories" in his (2008). There he distinguishes between (i) empirical fit and explanatory power, (ii) internal virtues, (iii) contextual virtues, and (iv) diachronic virtues (see McMullin 2008). And more recently, Michael Keas (forthcoming) provides yet another alternative taxonomy of the theoretical virtues. Keas identifies twelve theoretical virtues and groups them into four broad categories, which he calls evidential virtues, coherential virtues, aesthetic virtues, and diachronic virtues. Keas is less concerned with the relationship between the virtues and theoretical truth than he is with the relationship between the various theoretical virtues. To a large extent, his taxonomy builds on McMullin's, though he does disagree with McMullin on some issues.

[5] Duhem argues against the view that "*the aim of physical theory is to explain experimental laws*" (1906/1954, 10; emphasis in original). He argues that "a physical theory is ... a system of mathematical propositions, deduced from a small number of principles, which aim to represent as simply, as completely, and as exactly as possible a set of experimental laws" (19). Anti-realists often regard the explanatory power of a theory as of little evidential importance.

[6] Longino (1990) uses the term "constitutive values" in a broader sense than McMullin would. Longino counts "the values generated from an understanding of the goals of science" as constitutive, but she includes in this category "truth, accuracy, simplicity, predictability, and breadth" (see Longino 1990, 4). Larry Laudan also believes that we should count various values, like breadth of scope, as constitutive of science, even though he believes that it and other such values are "cognitive but non-epistemic values" (see Laudan 2004, 19). Such values, he argues, are not connected with the truth. I will follow McMullin's classification here, as it draws attention to important differences between the various values that Longino and Laudan group together.

themselves; they would not motivate us to carry on an activity in the first place" (68). But scientists certainly want theories that embody these values.[7]

Third, McMullin claims that there are other theoretical values that "are esteemed ... because they have proved to be the marks of a 'good' theory, a theory that will serve well in prediction and explanation" (1993, 68). Included in this class are "fertility, unifying power, and coherence" (68). This third class would also include simplicity. The connection between this class of values and the constitutive values of science is more tenuous than the connection between the second class and the constitutive values. A theory being inconsistent with other theories we accept clearly indicates that *something* is false, either the theory itself or some part of the other theories we accept. But when a theory fails to embody values in the third class, it might not necessarily be false. A theory that lacks simplicity, for example, may not be false, because the world may be quite complex in structure and thus defy simple conceptualization (see Cartwright 1983, 29).

McMullin argues that scientists have come to *learn* that theories embodying these latter values tend to serve the constitutive goals of science, that is, they tend to be better at prediction and explanation (1993, 68). But McMullin does not believe that the only justification for this third class of values is derived from past experience. He also believes that "these values *ought* to serve as indicators of a good theory. These are what one would *expect* a priori from a theory that purported to predict accurately and explain correctly" (68; emphasis in original). McMullin's evidence for this claim comes from the history of science. He notes that both Johannes Kepler and Robert Boyle "drew attention to the importance of such criteria, ... not to point to their

---

[7] I leave open the question of how extensive the compatibility must be between a particular theory and other theories a scientist accepts. Clearly, as a matter of *fact*, the various theories that scientists work with in different disciplines are not all consistent with one another. Kuhn came to describe the theories or lexicons in neighboring fields as incommensurable (see Kuhn 1991/2000, 98). This may not be as significant a problem as it might first appear to be, but it does sit uncomfortably with realism. Recognizing the limitations of our cognitive capacities, Christopher Cherniak argues that what we really ought to aim for is a "minimal consistency" (1986, § 1.5). Minimal consistency is a target somewhere between ideal consistency, which seems unachievable, and having no regard for logical consistency, which would be intolerable.

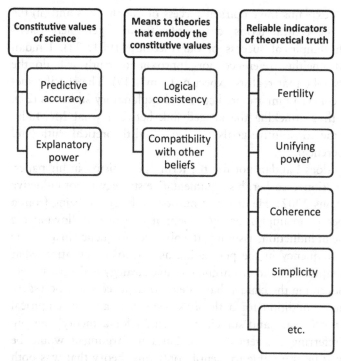

Figure 4 The relationships between the various theoretical values, according to McMullin

efficacy in the earlier history of natural philosophy but to recommend them on general epistemic grounds" (McMullin 1993, 68).

The relationship between the three classes of values is illustrated in the following diagram (Figure 4). McMullin provides a clear presentation of what realists typically assume about the theoretical values.

Laudan's Misunderstood Critique

As we saw in Chapter 5, Laudan's "Confutation of Convergent Realism" is frequently read as a defense of the Pessimistic Induction. This, I argued, is a mistake. The terms "pessimism" and "pessimistic" do not even appear in the article. Rather, the article is principally concerned with undermining various inferences that realists are inclined to draw from the success of our theories. Specifically, Laudan takes issue with the realists' claims that: (i) the approximate truth of our theories explains the empirical success of our theories and (ii) the fact that "the central terms in scientific theories genuinely refer" explains the empirical success of our theories (1981, 21). Laudan argues that "neither reference nor approximate truth will do the explanatory jobs that realists expect of them" (19). The specific successes that concern him are predictive and explanatory successes (21). Laudan is thus attacking the alleged link between what McMullin regards as the constitutive goals of science and theoretical truth (and genuine reference).

Not only does Laudan not use the term "pessimism" in his paper, but, as I mentioned earlier, his argumentative strategy is not inductive (see also Lyons 2002). He does not purport to be generalizing from a vast number of examples, as one might if one were following the straight rule of induction, assuming the observed frequency in a sample is the real frequency in the population as a whole. Repeating what I said in Chapter 5, Laudan's argumentative strategy is deductive. He aims to undermine the claims that (i) empirical success is a necessary and sufficient condition for a theory's being true and (ii) empirical success is a necessary and sufficient condition for a theory's having genuinely referring theoretical terms. Laudan's argument would be sound even if he were able to identify only one theory that was both successful and false, and only one theory that was both successful and did not have genuinely referring theoretical terms. Alternatively, he could achieve his goal if he could identify a theory that is approximately true but unsuccessful, and a theory that is unsuccessful despite the fact that it does have genuinely referring theoretical terms.

It is worth looking at this argument in detail. According to Laudan, contrary to what realists suggest, there have been numerous theories in the history of science that had genuinely referring theoretical terms but were unsuccessful (1981, 24). He cites as examples the chemical atomic theory in the eighteenth century and Wegener's theory of continental drift (24). Thus, Laudan concludes that "the realist's claim that we should expect referring theories to be empirically successful is simply false" (24). He also argues that there have been numerous successful theories whose central theoretical terms failed to refer (26–27). This claim is also supported by examples drawn from the history of science, "e.g., aether theories, [and] phlogistic theories" (25). Thus, a theory's having genuinely referring theoretical terms is neither necessary nor sufficient for it to be successful. Consequently, any inference from

(i) the predictive and explanatory successes of a theory to (ii) the claim that the entities postulated by the theory exist is dubious.

Laudan argues that the realists' claims about the connection between approximate truth and success are equally prone to refutation by evidence from the history of science. As he notes, "many theories which we believe to be false ... were – and still are – highly successful across a range of applications" (Laudan 1981, 30). He cites a number of examples, including "Newtonian mechanics, thermodynamics, [and] wave optics" (30). He also argues that theories can be approximately true but unsuccessful (see Laudan 1981, 30–32). He notes that there is no reason to think that the implications of a theory that is approximately true, that is, a theory that is partially false, will also be true. So, again, what Laudan has shown is that a theory's being approximately true is neither necessary nor sufficient for it to be empirically successful.[8]

The two key points Laudan wishes to establish in "Confutation" are that (i) there is no systematic connection between approximate truth and empirical success and (ii) there is no systematic connection between genuine reference and empirical success. But these sorts of connections are precisely what the realist needs to establish in order to support the sorts of inferences she makes on the basis of the success of our theories. And given the structure of Laudan's argument, it is futile for the realist to respond by attempting to show that any particular one of his examples mischaracterizes the situation. Nothing much depends on any single example.

Laudan (2004) extended his attack on the realists' appeal to the theoretical values, attacking specifically their appeal to breadth of scope, a value in McMullin's third class. Values in this class are alleged to be reliable proxies of the approximate truth of our theories. Laudan lists a number of different formulations that the appeal to scope has taken:

[1] acceptable theories are generally expected to explain the known facts in the domain;

---

[8] Laudan also has concerns about the notion of *approximate* truth (Laudan 1981, 31–32). Specifically, he does not think that realists have provided a clear explication of the notion of approximate truth. But even bracketing those concerns, he does not think the realist is warranted in making the sorts of inferences she makes from the success of our theories.

[2] acceptable theories are generally expected to ... explain different
kinds of facts;

[3] acceptable theories are generally expected to ... explain why their
rivals were successful; and

[4] acceptable theories are generally expected to ... capture their rivals
as limiting cases. (2004, 17; numerals added)

Laudan grants that "scientists frequently argue for one theory over
another if the former can explain or predict something about the world
not accounted for by its rivals" (2004, 17). Thus, he recognizes that, as
a matter of fact, scientists value theories with a broader scope. Laudan
even grants that "a theory is, all else being equal, better if it can explain
or predict facts from different domains or if it can show its rivals to be
limiting cases" (17). So he also thinks that there are good reasons for
valuing a theory that has a broad scope.

But Laudan claims that "none of these rules [[1]–[4], that appeal to
scope] can have an epistemic rationale since it is neither necessary nor
sufficient for the truth of a statement that it exhibit any of these
attributes" (2004, 18). Thus, though Laudan grants that a broad scope
is a genuine virtue of theories, he insists that it is not an *epistemic*
virtue. Scope, he claims, serves scientists' cognitive non-epistemic interests (19). But we have no reason to believe that a theory with a broad
scope is more likely true than one that does not have a broad scope.
And "no one has shown that any of [the] rules [[1]–[4]] is more likely
to pick out true theories than false ones" (18).

Realists may object that Laudan is attacking a straw-person realist.
After all, he believes that he has settled the issue, because he has shown
that scope is neither a necessary nor a sufficient condition for the truth
of a theory. Many contemporary realists would concede as much and
set their sights a little lower. They are not concerned with identifying
necessary and sufficient conditions for the truth of our theories.
Rather, in the spirit of a naturalized epistemology of science, they
would be content to show that theoretical values like scope are strongly
correlated with truth. Richard Boyd, for example, seems to hold such
a view (1980, 614). Realists who take a naturalistic turn claim that
the various theoretical values, like scope, are strongly correlated with
truth. Theories that have a broad scope are either more likely true than
theories that do not, or more likely true than not.

Though this sounds like a plausible response to Laudan's argument, the naturalists have not shown that even these weaker claims are supported by evidence gathered in a systematic way, in keeping with their naturalism. Rather, realists tend to take for granted the truth of the claims about the correlation between the various theoretical values and theoretical truth. Such claims are dogmas of realism. Indeed, realists have been quite critical of anti-realist appeals to the history of science, as we saw in Chapter 5, despite the fact that many of their own claims that play a crucial role in their arguments in support of realism depend on evidence from the history of science. So it seems that the realist is in as much need of evidence from the history of science as the anti-realist.

There is an alternative route that the realist might pursue in attempting to justify these sorts of methodological claims. The realist might take a Popperian attitude toward these methodological claims about the relationship between the theoretical values and theoretical truth. The realist might boldly hypothesize or conjecture such a link, and be open to rejecting it if the evidence suggests otherwise. But this Popperian type of defense is not compelling either. There have been many cases in the history of science where a simpler theory was inferior to a more complex theory or a theory broad in scope was inferior to a theory that was narrower in scope. Granted, confronted with such cases, the realist need not wholly reject her methodological claims. The realist could refine the claims, specifying, for example, under what specific conditions a simple theory is more likely true than a more complex theory. This revised methodological hypothesis would then be subjected to testing. But the realist has not pursued this strategy, at least not in any systematic way. So there is more work ahead for realists interested in appealing to the theoretical virtues. Though Laudan's arguments may only undermine a straw realist, the realists' own claims are not yet adequately supported either.

Theoretical Values and Ordinal Rankings

In the remainder of this chapter, I want to pursue a different argumentative strategy against the realist's appeal to the theoretical values. I will assume that Laudan is correct in showing that success is neither a necessary nor a sufficient condition for the approximate truth of our theories. Realists seem willing to concede this much. But for the sake of the argument, I will also grant to my realist opponents that there is a positive correlation between the various theoretical values, like simplicity and breadth of scope on the one hand, and theoretical truth on the other hand. I aim to show that even granting this much, the realist cannot get the evidential leverage she needs from the theoretical values. Confronted with a choice between two competing theories, the realist is not warranted in inferring that the theory that embodies the theoretical values to a higher degree is *approximately true*.

When a scientist makes an evaluation of the simplicity or scope of a theory, such a judgment is not a categorical judgment, like a judgment to the effect that a proposition is true or false. Rather, it is comparative. The scientist judges that one theory is *simpler* than another theory, or that one theory is *broader* in scope than another. On the basis of this sort of judgment, the scientist is not in a position to infer that the simpler theory is true, or even approximately true. All that such judgments yield is an ordinal ranking of the competing theories. When one theory is judged to be simpler than another, there is no fixed benchmark of simplicity against which this judgment can be measured.[9] So such evaluations, insofar as they support judgments about the truth or approximate truth of a theory, merely support the claim that one theory is *closer* to the truth than the other. But that is quite a different matter than inferring that the theory is *likely* true, or even approximately true. Indeed, from an evaluation of two competing theories with respect to their relative simplicity, we are in no position to know how far either is from the truth. All we can legitimately infer is that one theory is closer to the truth than the other. But both theories could be significantly far from the truth. This is similar to the point of Bas van Fraassen's Argument from a Bad Lot – when faced with a choice

---

[9] When scientists are trying to determine what curve best fits a series of data points, "simplicity can be given a reasonably precise meaning" (Steel 2010, 19). But often scientists are choosing between competing theories that cannot be adequately represented as two different curves through a set of data points. In such cases, simplicity is not such a straightforward notion. Despite the significant role that simplicity usually plays in discussions of the theoretical virtues, McMullin avoids much mention of it. He does, though, note that whereas some philosophers of science regard the simplicity of a theory as an indicator that it is likely true, others, like Cartwright, regard it as an indicator that the theory is likely false (see McMullin 2008, 503). Further, McMullin recognizes that realists have not yet explained *why* "a simple theory [is] more likely to be true than a less simple theory" (see McMullin 2008, 503).

between two theories, scientists might be choosing between theories that are both very far from the truth (van Fraassen 1989, 142–143). Similarly, when a theory is judged to embody the theoretical virtues more than any of its competitors, it may still be very far from the truth.

Importantly, it is not just the theoretical values in McMullin's third class that are unable to support inferences to the approximate truth of a theory. This argument applies equally well to the theoretical values in McMullin's first class, predictive accuracy and explanatory power. The fact that one theory is more accurate than another may give us reason to believe that it is closer to the truth than the other. But the greater accuracy of the theory does not warrant an inference to its approximate truth. And we are not in a position to determine how far it is from the truth, even when we know that it is more accurate than a competing theory.

Even some *realists* seem to recognize this limitation in theory evaluation. For example, Karl Popper (1963b) notes that

the status of truth in the objective sense ... and its role as a regulative principle, may be compared to that of a mountain peak usually wrapped in clouds. A climber may not merely have difficulties in getting there — he may not know when he gets there, because he may be unable to distinguish, in the clouds, between the main summit and a subsidiary peak. (1963b, 306)

Having reached a peak, one cannot infer that one has reached the summit. Realists, though, are typically assuming that every peak reached is the summit. But the theoretical values do not enable scientists to discern between subsidiary peaks and summits, that is, empirically successful but false theories and true theories.[10]

It is worth unpacking Popper's metaphor. The implication is that a theory can appear to be quite successful by a number of different

---

[10] Pavel Tichý (1974) provides an apt characterization of Popper's view. He describes his epistemological position "as an optimistic scepticism" (see Tichý 1974, 155). "It is a scepticism since it affirms that no non-trivial theory can be justified and that more likely than not all the theories we entertain and use are false" (155). What is odd about Popper's position is that he identifies as a realist. Ultimately, the reason he identifies as a realist is because he believes that scientists should *aim* for the truth. When they do not, as he thinks conventionalists and instrumentalists do not, they are likely to be satisfied with theories that may be effective at generating accurate predictions but are not true (see Popper 2002/1935, §§ 19–20). Scientists like Copernicus, Popper argues, achieved what they did because they were not satisfied with theories that merely generated accurate predictions.

measures but still may not be the true (or an approximately true) theory. Relying on the theoretical values as proxies or indicators, scientists can find themselves settling for a theory that is far from the summit they seek.

More recently, another realist, David Harker (2010), suggested that the realist should focus on the comparative success of competing theories and resist making any claims about *how* approximately true a particular theory is. Harker takes the comparative success of a theory to be a reliable indicator that it is closer to the truth than the alternatives to which it is compared (though not necessarily close to the truth). This much I have granted for the sake of the argument. Although they are realists, Harker and Popper do not presume that our best theories are true or even close to the truth. Theirs is a very modest form of realism. They seem to be more concerned with defending the notion of scientific progress. But this is not something anti-realists deny. The anti-realist, though, believes that the progress scientists make is with respect to their knowledge of the phenomena and does not necessarily extend to their knowledge of the unobservables.

Most realists who appeal to the theoretical values, though, are looking for more than this. In fact, most realists would not be satisfied with what either Popper or Harker delivers. Most realists appeal to the theoretical virtues in order to ground their claims that our current theories are likely true or approximately true.

Realism, Rationality, and the Problem of Theory Choice

In *Laws and Symmetry*, van Fraassen presents two conceptions of rationality, comparable to two different conceptions of the law, the Prussian conception and the English conception. "In the former, everything is forbidden which is not explicitly permitted, and in the latter, everything [is] permitted that is not explicitly forbidden" (1989, 171). We might call the conception that is similar to the English conception of law a *permissive* conception of rationality, as it allows anything that is not forbidden. And the conception that is similar to the Prussian conception of law we might refer to as a *directive* conception of rationality, as it explicitly tells us what we ought to do or believe.

Van Fraassen endorses the permissive conception. He argues that "it is rational to believe … anything that one is not rationally compelled to disbelieve" (1989, 171–172). He thus asserts that *"rationality is only bridled irrationality"* (172). Given a permissive conception of rationality, it is rational to believe anything that does not conflict with the canons of rationality. Importantly, though, we are not required to believe everything that is permitted by the canons of rationality. We are only prohibited from believing what conflicts with the canons. So if the choice between two theories is underdetermined by the evidence, a scientist can rationally (i) believe one theory, or (ii) believe the other theory, or (iii) believe neither theory. And as far as van Fraassen is concerned, there is no reason to *believe* any theory. It is enough to merely accept a theory to work with it (1980, 12). Moreover, given the permissive conception of rationality, there is nothing irrational about believing a theory either, provided such belief does not conflict with the canons of rationality. Such beliefs, though, merely expose one to a greater risk of holding false beliefs, a risk that van Fraassen does not want to take.

The realist, on the other hand, seems to assume a directive conception of rationality. This is evident from the fact that when the realist asks which theory embodies the theoretical values to the highest degree, she aims to *determine* which theory she (and *everyone else*) ought to believe. The assumption is that it would be irrational to believe any theory other than the one that is ranked the highest according to the theoretical values. We want the best theory we can have, the realist claims, and the theoretical values are our best, perhaps *only*, means of determining which theory that is. This is why it is so important for the realist to insist that these values are strongly correlated with theoretical truth. Were they not, they would be inadequate guides for theory choice.

Some realists seem to go even further and suggest that it is *irrational* not to believe that a theory is true or approximately true if it is deemed superior as measured by the theoretical values. Withholding belief, they claim, amounts to an untenable skepticism (see Lipton 1993/ 1996). But as we saw above, the theoretical values are inadequate guides for theory choice. They do not provide warrant for such judgments. At best, they are reliable, though fallible, indicators of whether one theory is closer to the truth than another theory. But this will not do, given the realist's goals. After all, one of two competing theories can be closer to the truth than the other even when both theories are quite far from the truth. The fact that the theoretical values merely aid scientists in generating an ordinal ranking of competing theories undermines the warrant for an inference from the success of a theory to its truth or approximate truth.

Directive Accounts of Rationality Are Shortsighted

Things are actually more complicated than I have suggested in ways that further frustrate the realist's appeals to the theoretical values. The realist seems to assume that the rational action is for scientists to work with the theory that is closest to the truth, the simpler theory, for example, or the one broader in scope, or, ideally, the one that embodies all the theoretical virtues to the highest degree. I aim to show that even this is a fallacious inference.

Even after it is determined that one theory is simpler than another theory, and therefore closer to the truth, one cannot infer that accepting and working with the simpler theory is likely to lead scientists to the truth or even closer to the truth in the long run than working with the more complex theory, even if we grant that the theory judged to be simpler is closer to the truth than the more complex theory. The problem is that there may be particular features of the simpler theory that will prove to be impediments to *further improvements*. Once scientists recognize that they are working with imperfect theories, they should realize that working with a theory that is closer to the truth is not necessarily the best path to follow in order to get a better theory (one even closer to the truth, as the realist would have scientists aim to do). Working with the more complex theory may be the more expedient path to a better theory.

Recall Popper's analogy, cited above. In the language of Popper's analogy, a scientist may get caught on a subsidiary peak, unable to reach the summit. Consider a case where scientists are choosing between two competing theories, T1 and T2, where T2 is deemed to be (and in fact is) closer to the truth as determined by the theoretical values. In some situations, choosing to work with a particular scientific theory, T2 rather than T1, may prevent scientists from getting to an even better theory, T3. Features of the theory deemed to be superior, T2, may prove to be serious impediments to future improvements in a way that comparable features of its weaker competitor, T1, are not. Guided by a directive conception of rationality, though, scientists will be reluctant to work with a theory that is further from the truth, that is, T1, than some existing competitor that is closer to the truth, T2.

Realists, committed to a directive conception of rationality, are thus at risk of binding themselves in counterproductive ways. Insofar as the realist assumes a directive conception of rationality, the realist's advice to scientists will be to work with the theory that embodies the theoretical values to the higher degree. But this strategy will make scientists vulnerable to getting caught on subsidiary peaks. Importantly, this concern is not just a new version of the argument from the underdetermination of theory choice by data. The concern here is that the path to ever-better theories is not necessarily either linear or progressive. Consequently, we cannot assume that choosing to work with the theory closest to the truth will get us to the truth in the long run. Nor is the concern I raise the same as Kuhn's (1977) concern, that different scientists may be led to make different evaluations of competing theories. The concern I am raising would arise even if there were unanimous agreement about which of two competing theories is superior, and even if the superior theory were closer to the truth than its competitor. The concern I am raising is that a commitment to the superior theory may in fact impede scientists in developing an even better theory.

Thus, the theoretical values fail to both (i) provide warrant for the inferences that realists want to draw and (ii) solve the practical problem of determining which theories scientists ought to work with.

We are now in a position to determine what support the theoretical values *really* offer the scientific realist. The short answer is not much. My arguments suggest that realists misunderstand the role that theoretical values *should* play in science. When scientists appeal to the theoretical values to evaluate competing theories, what they get is merely an ordinal ranking of the theories, not the sort of ranking that can support an inference to the truth or likely truth of the superior theory. Further, if scientists are aiming to develop better theories given a set of theories to choose from, working with the theory that embodies the theoretical values to the greatest extent may not be the most expedient way to develop better theories.

It is worth distinguishing the point I am making about theory evaluation in this chapter from the point I made in Chapter 3, when I defended the Argument from Underconsideration. In this chapter, I argued that when scientists appeal to the theoretical values in their efforts to evaluate competing theories, they only yield ordinal rankings. An ordinal ranking, though, cannot support an inference to the truth of a theory. In my analysis and defense of the Argument from Underconsideration, the problem that I highlighted is the fact that theory evaluation is comparative. Comparative evaluations also undermine any sort of warranted inference to the truth of a theory. But they do so because scientists may be choosing between a pair of theories, neither of which is true or approximately true. These are two distinct concerns that threaten the realists' inference to the truth of our best theories. Common to both arguments is the concern that knowing that one theory is superior to another does not warrant an inference to the likely truth or approximate truth of the superior theory. Neither comparative evaluations nor ordinal rankings of theories give us information about the relative distance of our theories from the truth.

With respect to the practical problem raised above, it seems that the anti-realists' permissive conception of rationality puts scientists in a better position than the directive conception of rationality. By permitting a scientist to do whatever is not contrary to the canons of rationality, the anti-realist permits different scientists to make different choices, which may in fact better serve the goals of the research community as a whole.

PART II

Strengthening Anti-Realism

9 | But Can the Anti-Realist Explain the Success of Science?

So far I have concentrated on arguments against realism. In the remainder of the book, I want to focus on the strengths of anti-realism. I want to examine two issues. First, I want to examine how the anti-realist can explain the success of our scientific theories without assuming that they accurately represent the underlying structure of reality. Second, I want to examine the insight that the anti-realist can offer into why even our best contemporary theories are apt to be replaced in the future by theories that make significantly different assumptions about the structure of reality.

Realists claim that they have one important advantage in the debate with anti-realists. They claim that, unlike the anti-realists, they have an explanation for the success of science. Indeed, this is the key point in the No Miracles Argument for scientific realism. The realists' explanation, the appeal to the truth or approximate truth of our theories, they claim, is the only plausible explanation for the success of our current best theories. Were our successful theories not true or at least approximately true, their success would be inexplicable, if not miraculous.

Contrary to what realists suggest, anti-realists have developed an explanation for the success of science. Specifically, Bas van Fraassen has developed, albeit in a sketchy form, a selectionist explanation for the success of science, an explanation modeled on selectionist explanations in biology. My aim in this chapter is to defend van Fraassen's selectionist explanation for the success of science. In the process, I will also provide grounds for rejecting the realists' No Miracles Argument.

First, I examine the realists' No Miracles Argument. Then I present van Fraassen's criticism of the argument and his alternative explanation for the success of science, his selectionist explanation. I then argue that van Fraassen's explanation is superior to the realists' explanation in two respects. First, unlike the competitor explanation, it can explain why it is that we come to reject widely accepted theories that were regarded as successful in the past. Second, unlike the competitor explanation, van Fraassen's selectionist explanation can account for the fact that sometimes two competing theories are both successful. Finally, I address the concern that only realists can explain the predictive success of novel phenomena.

The No Miracles Argument

Let us begin with an examination of the No Miracles Argument. Realists often claim that it is *because* our theories accurately reflect the structure of the world that they routinely enable us to make accurate predictions. Hilary Putnam presents a version of this argument, arguing that, given the success of our current theories in mature fields, it would be a miracle if our theories did not accurately reflect the structure of the world (see Putnam 1975, 73). Clearly, no one believes that the success of our current theories is due to a miracle. Consequently, realists argue that the best explanation for the predictive success of our current theories is that they do in fact accurately reflect the structure of the world.

A similar argument was developed by J. J. C. Smart in the early 1960s. Rather than suggesting that if our theories are not true (or at least approximately true), then the success of science is due to a miracle, Smart suggests that if our theories are not true, then the success of our current theories is due to some sort of cosmic coincidence (see Smart 1963/2009, 39). The implication here is that if our theories genuinely misrepresent the unobservable structure of the world, then it is just some sort of lucky coincidence that our theories, despite being false, continue to generate true predictions. Smart claims that the success of merely instrumentalist theories would seem to depend upon a cosmic coincidence (Smart 1963, 39). *Merely* instrumentalist theories are theories that were designed with no intention of accurately representing the underlying structure of the world. Instead, such theories are designed with the intention of merely saving the phenomena, accounting for the observables, and making true predictions. It would be less embarrassing to attribute the success of science to a cosmic coincidence than it would be to attribute it to a miracle, but both of these explanations are highly implausible. Cosmic coincidences are, after all, assumed to be highly improbable events, perhaps only slightly more probable than miracles.

It is worth noting that Smart's motives for presenting this argument are somewhat different from Putnam's motives. Smart presents his argument, the Cosmic Coincidence Argument, in a discussion of the implausibility of the instrumentalist view of scientific theories. Instrumentalists claim that scientific theories are not intended to be representations of the world. Rather, they are merely instruments, designed to enable scientists to generate accurate predictions of the observables.[1] When Smart was writing, the type of instrumentalism that concerned him was associated with "phenomenalism about sub-microscopic objects." According to this view, "sentences about electrons, protons, and the like can be *translated* into sentences about galvanometers, cloud chambers, and the like. On this view electrons and protons are logical constructions out of macroscopic objects" (see Smart 1963, 27).

This view was held or allegedly held by a number of influential philosophers of science and scientists at the turn of the twentieth century, including Ernst Mach (1897/1984) and William James (1907/1949). Karl Popper claims that, in addition to Mach, "Kirchhoff, Hertz, Duhem, Poincaré, Bridgman, and Eddington ... [were] all instrumentalists in various ways" (see Popper 1956/1963a, 133, note 5). The early Logical Positivists, especially in the Vienna Circle days, were also sympathetic to instrumentalism (see Popper 1956/1963a, 145). But this seems to have changed by the early 1960s, when the Logical Positivists and Logical Empiricists were based mainly in the United States and when Smart was developing his Cosmic Coincidence Argument.

By the time Putnam was writing on the realism/anti-realism debate, the concern was no longer with instrumentalism, but rather with the question of whether we had good reason to believe that our theories

---

[1] Writing around the same time, Karl Popper (1956/1963a) also took issue with instrumentalist accounts of science. According to Popper, "the instrumentalist view asserts that theories are *nothing but* instruments" (1956/1963a, 136). He suggests that the instrumentalists regard "science [as] nothing more than glorified plumbing, glorified gadget making" (137) and theories "are nothing but computational rules (or inference rules)" (149). Interestingly, Popper suggests that "the *instrumentalist view* ... has become an accepted dogma. It may well be called the 'official view' of physical theory since it is accepted by most ... theorists of physics" (134). His concern with instrumentalism is somewhat different from Smart's concern. Popper claims that the instrumentalist is "unable to account for real tests, which are attempted refutations" (1956/1963a, 151–152).

are true or we should be skeptical of the possibility of theoretical knowledge. Putnam presents the No Miracles Argument in a discussion of the strongest arguments for and against scientific realism.

Let us consider the structure of the No Miracles Argument:

P1. Our current theories in mature fields routinely enable us to make accurate predictions. That is, our current theories are successful.

P2. The success of our current theories is due either to the fact that they accurately represent the world or to a miracle.

P3. The success of our theories is not due to a miracle.

C.   Consequently, the success of our theories is due to the fact that they accurately reflect the structure of the world. Consequently, our theories accurately reflect the structure of the world.[2]

In an effort to strengthen the argument, Alan Musgrave has proposed two amendments to it (see Musgrave 1988). As far as he is concerned, it is not enough just to show that the realists' explanation for the success of science is better than the "miracle" explanation. As Musgrave explains,

the fact to be explained is the (novel) predictive success of science. And the claim is that realism ... *explains* the fact, explains it *satisfactorily*, and explains it *better* than any non-realist philosophy of science. And the conclusion is that it is reasonable to accept scientific realism ... as true. (1988, 239; emphasis in original)

First, Musgrave concedes that it is not enough that a particular explanation is the best explanation we have managed to develop for the success of science. In addition, an adequate explanation for the success of science, one worthy of our acceptance, must be *satisfactory*. So there is some absolute minimal standard, albeit hitherto unspecified, that an acceptable explanation for the success of science must meet. Second, Musgrave insists that it is the successful predictions of *novel* phenomena that need explaining. When a theory routinely predicts phenomena similar to the sorts of phenomena that it was designed to predict, such

---

[2] Sometimes the realists' argument is laid out as an instance of the fallacy of affirming the consequent (see Brown 1985; Musgrave 1988). Assuming the truth of a theory, an observation is predicted. And the prediction is confirmed by experience, thus leading to the unwarranted conclusion that the theory is true (see also Laudan 1981, 45). The advantage of the way I have presented the realists' argument is that it is at least a valid inference. It is an instance of disjunctive syllogism.

predictions do not provide strong support in favor of either a specific theory or the realists' explanation for the success of science. But predicting a phenomenon that the theory was not explicitly designed to predict is impressive. It is then that the success of a theory seems like a miracle or a cosmic coincidence if the theory is not true or at least approximately true.

There is much debate about precisely what types of predictions constitute predictions of novel phenomena. For example, some argue that a predicted phenomenon counts as novel provided it was not used in constructing the theory, even if scientists knew about it (see, for example, Musgrave 1988, 232; Leplin 1997, 50–51). This has come to be called "use-novelty." Others claim that temporal novelty is the salient notion. That is, a phenomenon counts as novel provided scientists did not know about it "before they derived it from a [theory]" (see Leplin 1997, 41). And other definitions of "novel predictions" have been proposed (see, for example, Alai 2014, 297). We need not resolve this debate about novel predictions now, as none of the concerns I raise in the remainder of the chapter depends on settling this issue one way or another. The key point is that vindicated predictions of novel phenomena seem especially challenging for the anti-realist to explain.

Van Fraassen's Assessment of the Realists' Explanation

Van Fraassen is deeply suspicious of both the realists' explanation for the success of science and the No Miracles Argument. Let us consider his criticism of the argument first. Like the realists, van Fraassen is impressed by the success of our current theories. He grants that our current best theories in mature fields are capable of yielding accurate predictions on a routine basis. Further, like the realists, he believes that this is no miracle (1980, 40). What van Fraassen objects to is the choice offered by the realist: either (i) the success of science is a miracle or (ii) our current theories accurately represent the structure of the world. That is, he believes that the second premise of the argument is false. The choice is presented as an exhaustive dichotomy, but in fact it is not. Van Fraassen believes that there is a third viable alternative explanation. Most importantly, he does not believe that the *best* explanation for the predictive success of our current theories is that they accurately reflect the structure of the world. Instead, he attributes the success of our current theories to the fact that unsuccessful theories have been eliminated in a process of selection comparable to the selection process operative in the biological world.

Let us consider van Fraassen's selectionist explanation for the success of science. But first let us consider a typical selectionist explanation in biology. Darwin tells us that the best explanation for the remarkable fit between biological organisms and the environments they inhabit is that natural selection ensures that organisms lacking such a fit are destroyed (see Darwin 1859/2003). Seals, for example, have flippers that are effective for swimming, but not because they are intentionally designed for this purpose. Rather, those seals that have the misfortune to be born with ineffective flippers are more apt to die before they are able to reproduce. And even if they live long enough to reproduce, they are likely to produce fewer offspring than those born with effective flippers, as the less effective flippers expose them to greater dangers every day. Seals with variants of ill-formed flippers are thus weeded out of the population. Consequently, when we look at the biological world, we tend to find only organisms that fit the environment they inhabit. The seal is no exception in this respect.

Van Fraassen offers a similar explanation for the success of our current theories. As he explains, any theory that does not enable scientists to make accurate predictions is not apt to be around very long. No scientist will waste her career working with such a theory. As a result, any theory that is still around, that is, any theory that is still being used by scientists, is apt to be successful. Consequently, when philosophers of science look at the world of science, they should not be surprised to find only successful theories. The others have been eliminated or are on their way to being eliminated.

Van Fraassen notes an additional similarity between theories that retain a following and organisms that continue to survive. He compares successful theories to mice that run from cats. Van Fraassen notes that all the mice we encounter run from cats. But in order to explain this behavior, we need not and should not assume that mice accurately represent their environment. We need not even assume that the mouse perceives the cat to be an enemy (van Fraassen 1980, 39). Such an explanation embodies the presumptions of the realists. Instead, van Fraassen claims that all we need to assume is that those mice that did not acquire the habit of running from cats are no longer with us. Thus, the current success of mice can be explained without recourse to questionable assumptions or conjectures about the underlying thoughts of mice, assumptions and conjectures that may or may not be true. Similarly, the success of our current theories can be explained without recourse to questionable assumptions about our success in latching on to the unobservable structure of the world. Our theories enable us to make accurate predictions, because scientists do not work with theories that do not enable them to make such predictions. Consequently, inaccurate theories are not represented in the population of theories accepted by scientists. Scientists who work with unsuccessful theories are as rare as mice that do not run from cats, and the fate of both is similar.

Van Fraassen clearly shows that the choices presented in the second premise of the No Miracles Argument are not exhaustive. The ease with which van Fraassen devises an alternative explanation for the success of science suggests that the choice presented by the realists is not forced upon us. Consequently, we should be reticent to accept the conclusion of the No Miracles Argument, given that one of the premises is so questionable. Certainly, before we accept the realists' explanation for the success of science, a better argument is required.

The Key Strengths of Van Fraassen's Explanation

Both the realist and van Fraassen offer *plausible* explanations for why our current theories enable us to make accurate predictions.[3] Hence, in an effort to distinguish which explanation is superior, we will need to look beyond their explanations of the fact that we have theories that are predictively accurate.

In this section, I argue that van Fraassen's selectionist explanation is superior to the realists' explanation for two reasons. First, unlike the realist, van Fraassen can explain why long-accepted theories come to be rejected without having to retract an earlier explanation for the success of the now rejected theories. Second, unlike the realist, van Fraassen can explain why two competing theories can both be predictively successful.

Unlike the realists' explanation, van Fraassen's explanation for the success of science enables us to explain why long-accepted theories

---

[3] Some *realists* explicitly acknowledge the limitations of the No Miracles Argument, that is, the fact that it is, at best, a plausible explanation for the success of science (see, for example, Brown 1985, 66; Musgrave 1988, 249).

come to be rejected. That is, it provides us with the resources to explain the *failures* of science. The realist seems to have nothing to say here. The realist cannot say in good conscience that a theory that once reflected the structure of the world no longer does so. That is contrary to our common understanding of what is involved in a theory reflecting the structure of the world. The realist, after all, assumes that the underlying structure that scientists aim to get at with their theories is essentially unchanging.[4]

I anticipate that realists may object to my characterization of the situation and claim that they can reconcile the rejection of past successful theories with their claim that our current predictively accurate theories are likely approximately true. I anticipate two sorts of replies from realists to the concern I raise. First, some realists are apt to emphasize our fallibility. This is Ilkka Niiniluoto's (1999) strategy. As Niiniluoto explains, given that the realist is a fallibilist, she

may admit counter-examples — in the same way as a rule of induction may rationally assign a high epistemic probability to a generalization on the basis of repeated observational successes, but the generalization after all turns out to be false. (Niiniluoto 1999, 192)

Niiniluoto is right to emphasize human fallibility. This seems to be a reasonable part of any naturalized epistemology of science. But this appeal to scientists' fallibility does not constitute an adequate reply to the concern I raised above. It makes matters a little too easy for the realist. It seems that the realist is apt (or may be tempted) to regard each theory that is replaced by a better successor theory as an admissible counterexample. This explanatory strategy is suspect, for it seems that the truth (or approximate truth), the explanans of the realist's explanation, does not explain the predictive success of the now-rejected but previously accepted successful theory.[5] It also seems like

---

[4] Alexander Bird puts the point in the following way: "the features of the world [our theories] respond to are what they are independently of our theories, and are by and large constant over time" (Bird 2000, 213). Michael Devitt also notes that realists emphasize the mind independence of the "unobservables of well-established current scientific theories" (see Devitt 2014, 257). Oddly, at least in the case of Devitt, the implication is that the anti-realist believes otherwise.

[5] Philip Kitcher (1993) and Stathis Psillos (1999) suggest that we distinguish between those parts of a theory that are responsible for its success and those parts that are not. They claim that it is only the latter that are subject to change with a change of theory. The former are not, and they constitute the true parts of the

an ad hoc way of dealing with the failure of theories. The realist, it seems, will stick to her realist explanation for any particular theory until it is replaced. Then the realist will appeal to the fallibility of scientists, and insist that this particular instance is an exception.[6]

The second response that I anticipate from the realist is as follows. The realist will note that there is a pattern in the process of scientific change, a pattern that emerges over long periods of time. Successor theories are more successful than their predecessors, and these more successful successor theories are superseded by even more successful theories. This fact, the realist claims, suggests that with each successive theory in a field, scientists are getting ever closer to the truth (see, for example, Kitcher 1993, 150–151; also Harker 2010). The increasing accuracy of the predictions made by scientists seems to support this line of reasoning. Given this line of reasoning, the realist claims that once-successful theories are rejected because they have been superseded by theories that *better* represent the unobservable structure of the world. So the apparent failures of science are actually just part of the success of science.

But even this explanatory strategy, this appeal to convergence, has its problems. As Larry Laudan (1981) argues, as a historical matter of fact, scientists have not preserved the mechanisms, models, and laws of earlier theories through theory change. Laudan provides a catalogue of examples to support his claim:

Copernican astronomy did not retain all the key mechanisms of Ptolemaic astronomy (e.g., motion along an equant); Newton's physics did not retain all (or even most) of the 'theoretical laws' of Cartesian mechanics, astronomy and optics; Franklin's electrical theory did not contain its predecessor (Nollet's) as a limiting case. Relativistic physics did not retain the aether, nor the mechanisms associated with it; statistical mechanics does not incorporate all the mechanisms of thermodynamics; modern genetics does not have

---

theory. This "selective realism," as it has come to be called, has been criticized for offering only a post hoc explanation for the success of our theories (see Stanford 2003, 569; Chakravartty 2007, 46). Psillos has addressed this criticism, insisting that we can "independently identify the theoretical constituents that contribute to the successes of a given theory" (see Psillos 1999, 112). He argues that "eminent scientists do the required identification all the time" (Psillos 1999, 112).

[6] Karl Popper explicitly objects to meeting challenges to our theories by making ad hoc adjustments, that is, by *accommodating* recalcitrant evidence (see Popper 1935/2002, 64). Insofar as scientific realism is itself a scientific theory, Popper would not condone such a response.

Darwinian pangenesis as a limiting case; the wave theory of light did not appropriate mechanisms of corpuscular optics; modern embryology incorporates few of the mechanisms prominent in classical embryological theory. (Laudan 1981, 39)

So, contrary to what many realists assume, the history of science does not support the claim that with each change of theory, scientists are converging on the truth. Were scientists in fact getting ever closer to the truth with each change of theory, we would expect the successor theories to retain the successes of the theories they replace. But this is not so.

Laudan also argues that from a normative point of view, convergence is not even desirable. According to him, if scientists were to commit to only accepting new theories that retain *all* the successes of the theories they replaced, it would put unwarranted, and potentially unproductive, constraints on science. As Laudan notes, "some of the most important theoretical innovations have been due to a willingness of scientists to violate the cumulationist or retentionist constraint which realists enjoin ... scientists to follow" (Laudan 1981, 39; see also Feyerabend 1988). Indeed, as we saw in Chapter 5, even Nicholas Rescher, a committed realist, insists that we should not identify scientific progress with convergence. Radical theory change is incompatible with convergentism (Rescher 1987, 24–25). And given the historical record, radical theory change seems to be an undeniable part of the growth of science (Rescher 1987, 15; Worrall 1989).

Unlike the realist, the selectionist can readily provide an explanation for the rejection of once-successful theories. Consider the situation of mice again. If mice find themselves in an environment with a new predator, we may find that the disposition to run from cats will no longer explain their survival. That is, having the disposition to run from cats is no longer sufficient to explain the survival of mice. The new predator may thus create a *new standard* of success, and this will need to figure in an adequate explanation for the survival of the then-successful mice. The python, an invasive species in the Florida Everglades, for example, has radically changed the challenges that the indigenous species must overcome to ensure their survival.

The situation is similar in science. As a field develops, theories can be expected to explain things they were not expected to explain in the past. Our theories thus face new challenges, including some they were not initially designed to address. This is one reason why very successful theories can come to be regarded as unsuccessful, and thus come to be rejected. Sometimes the currently accepted theory will be able to meet new challenges. But sometimes the currently accepted theory will not be able to meet the new challenges it encounters.

Let us consider a concrete example from the history of astronomy. After Galileo observed the moons of Jupiter, *all* astronomers were expected to explain how satellites remain in their orbit around an orbiting planet. Before Galileo reported his observations of Jupiter's moons, this was not a problem for Ptolemaic astronomers. To some extent, it was a problem internal to the Copernican theory.[7] According to the Copernican theory, the moon orbits the Earth as the Earth orbits the sun. In the Ptolemaic theory, every celestial body orbits the Earth and the Earth is stable at the center of the cosmos. Consequently, in the Ptolemaic theory, the problem of explaining how a satellite remains in its orbit never arises. According to the Ptolemaic theory, there are no such entities. But after Galileo's telescopic observations of the moons of Jupiter, new successes were expected of *all* theories. Even the Ptolemaic astronomer had to explain how Jupiter's moons keep up with the planet as it (allegedly) moves around the Earth. Though this did not constitute a decisive strike against the Ptolemaic theory, it did eliminate an advantage that it had over the Copernican theory.[8]

Galileo's discovery of the phases of Venus had an even more devastating effect on the Ptolemaic theory. But in this case, Galileo's observations did not just eliminate an advantage of the dominant theory. Rather, as we saw in Chapter 1, the Ptolemaic theory was incompatible with the new data. Thus, in this case, the new challenge contributed significantly to the ultimate defeat of the Ptolemaic theory. The new challenge it faced was more than it could bear.

The fact that once-successful theories come to be rejected is something that van Fraassen's explanation for the success of science seems better suited to explain. The realist, on the other hand, leaves us with no explanation. All the realist can say is that the theory that scientists *thought* reflected the structure of the world does not. But then the

---

[7] Advocates of Tycho Brahe's theory of planetary motion were faced with a similar problem. In Brahe's theory, the planets orbit the Sun, which orbits the Earth.

[8] Toulmin also emphasizes, in his evolutionary model of scientific change, that the standards by which theories and contributions are judged are determined by various historical contingencies (see Toulmin 1981, 27).

theory's prior success cannot and never could be explained by the fact that it reflects the structure of the world.[9]

Let us now consider the second way in which van Fraassen's selectionist explanation is superior to the realist's explanation. Unlike the realist's explanation, van Fraassen's account of scientific success can explain why two competing and contradictory theories are both predictively accurate. Again, let us consider a concrete example from the history of astronomy. As we saw in Chapter 1, during the late 1500s, Copernicus' theory of planetary motion and the late Renaissance version of the Ptolemaic theory of planetary motion were both predictively accurate. In fact, the two theories were roughly equally accurate with respect to the predictions they generated, erring by as much as 5 degrees with respect to some predictions, but often predicting with far greater accuracy (see Gingerich 1971/1993; 1975b/1993; Thoren 1967).

Realists seem to have nothing insightful to say about such situations. They certainly cannot pursue their standard explanatory strategy, attributing the predictive success of the two theories to the fact that *both* theories accurately represent the underlying structure of the world. After all, the two theories compete, and, interpreted *literally*, they ascribe a different structure to the world. Most importantly, Ptolemaic astronomers assumed that the Sun orbits the Earth, whereas Copernicans claimed that the Earth orbits the Sun.

It is doubtful that *anyone* wants to attribute the predictive success of the Ptolemaic theory to the fact that it accurately represents the structure of the world.[10] But, as noted in Chapter 1, in order to develop a theory that was as successful predictively as the contemporary Ptolemaic theory, Copernicus had to invoke epicycles and deferent circles, irregularities that he regarded as misrepresenting the structure of the solar system. Hence, even granting that the Copernican theory is a

---

[9] This fact, that van Fraassen's selectionist explanation provides an explanation for both the successes and failures of science, should appeal to proponents of the strong programme in the sociology of scientific knowledge. Their symmetry principle demands that the same causes explain both the successes and failures of scientists (see Barnes and Bloor 1982, 22–23).

[10] Surprisingly, in his discussion of the success of the Ptolemaic theory, Niiniluoto (1999) claims that, before the development of Copernicus' theory, "it was indeed rational to regard Ptolemy's well-developed theory as the most truthlike of the existing astronomical theories" (192). It is unclear exactly why Niiniluoto makes this claim.

more accurate representation of reality insofar as it acknowledges that it is the Earth and other planets that orbit the Sun, the theory's predictive success was not a consequence of the fact that it mirrors reality. Rather, its predictive success was a consequence of the fact that it employs eccentric circles, epicycles, and deferent circles. These were strategically and deliberately built into the planetary models to ensure that the theory was as successful as the contemporary Ptolemaic theory. In short, these were ad hoc adjustments.

The fact that two competing theories are both predictively accurate is not a problem for van Fraassen's account of scientific success. When two competing theories both enable scientists to make accurate predictions of observable phenomena, we should expect each theory to be accepted by some scientists. Indeed, this is what we find in the history of astronomy, as long as the two competing theories remained roughly equally accurate. Thus, unlike the realist, van Fraassen can explain the successes of science even when the successes are divided between competing theories. Thus, there is good reason to believe that the predictive success of *some* of our theories is due to something other than the fact that these theories accurately represent the structure of the world.

Predictions of Novel Phenomena

There is one issue that realists are especially inclined to emphasize in their explanation for the success of science. Realists often claim that, unlike the anti-realists, they can explain why our successful theories enable us to make *novel* predictions. We saw this earlier, in Musgrave's amended version of the No Miracles Argument. Such successes are not surprising if, as the realist claims, our successful theories accurately represent the structure of the world. But the realist argues that such successes really are miraculous if our theories do not accurately represent the structure of the world (see Musgrave 1988; Leplin 1997).[11]

But even here the realist is mistaken. A number of philosophers of science have noted that scientists have frequently generated vindicated predications of novel phenomena from false theories. Martin Carrier,

---

[11] Van Fraassen does not discuss the distinction between successful predictions and successful novel predictions.

for example, cites two cases of false theories that generated true predictions of novel phenomena, specifically, "Priestley's prediction of the reductive properties of hydrogen based on the phlogiston theory ... [and] Dalton's ... prediction of the equality of thermal expansion of all gases based on the caloric theory of heat" (1991, 29). Both the phlogiston and caloric theories of heat make assumptions about reality that we now regard as false; false, that is, by the lights of the theories we accept today. Importantly, as Carrier explains, these are cases "in which wrong aspects of wrong theories are responsible for ... [novel] predictive success" (1991, 29). So these theories seem to be immune to Stathis Psillos's *divide et impera* strategy (see Psillos 1999, 108–114). Assuming Carrier is correct about the history, the realist would be led to treat as truth-like features of these theories that we now regard as false.

Importantly, these are not just isolated cases where false theories generated true predictions of novel phenomena. Both Timothy Lyons (2002; 2006; 2012; forthcoming) and Peter Vickers (2013) identify numerous theories in the history of science that are now regarded as false that were able to generate predictions of novel phenomena, that is, phenomena they were not initially designed to account for. These include the caloric theory, Newtonian mechanics, Fresnel's wave theory of light, Dalton's atomic theory, Mendeleev's Periodic Law, and Bohr's theory of the atom (see Lyons 2002, 70–72). In some cases, the false theories generated a number of predictions of novel phenomena. Consequently, contrary to what the realist suggests, we cannot even take the fact that a theory enables us to make a true novel prediction as a reliable indicator that the theory is true or likely true.

Granted, the realist may be correct to suggest that it is *mysterious* that false theories are able to make successful novel predictions. But the facts are as they are. Theories that misrepresent the world can enable and sometimes have enabled scientists to generate vindicated predictions of novel phenomena. Hence, contrary to what realists imply, novel predictions do not settle the case in favor of realism.

Indeed, the shortcomings with the No Miracles Argument that I have discussed are not the only ones that speak against it. Greg Frost Arnold (2010), for example, has argued that the sort of scientific realism that the No Miracles Argument defends, though purporting to be a *scientific* explanation for the success of science, fails to measure

up to the standards of science. For example, it makes no new predic-
tions, as we would expect from a scientific explanation (see also
Doppelt 2005, 1080). And Magnus and Callender argue that the No
Miracles Argument commits the base rate fallacy (see Magnus and
Callender 2004, §§2–4; see also Howson 2000, 54; Lipton 2004,
197–198; Worrall 2012, § 4.3).

In summary, I have argued that the realists' No Miracles Argument
should not persuade us. It depends on a false premise. We are not
forced to attribute the success of our theories to either the fact that they
are true or the fact that a miracle has ensured their success. Further,
I have argued that van Fraassen's selectionist explanation for the
success of science is superior to the competitor realist explanation.
Though the realist and van Fraassen both offer *plausible* explanations
for why our current theories enable us to make accurate predictions,
only van Fraassen's selectionist explanation provides us with resources
to adequately explain the failure of past successful theories and the fact
that two competing theories can both be successful.

10 | Selection and Predictive Success

In the previous chapter, we examined the realists' No Miracles Argument and Bas van Fraassen's selectionist explanation for the predictive success of our current best theories. In this chapter, I want to examine the viability of a selectionist explanation for the success of science in more detail. Scientific realists have criticized van Fraassen's selectionist explanation for the success of science quite extensively. And some of the criticisms seem to be based on misunderstandings of the nature and purpose of the explanation. I aim to clarify the nature of the selectionist explanation for the predictive success of our best theories and defend it from a variety of common criticisms.

Surprisingly, van Fraassen actually says very little about his selectionist explanation. Indeed, what he says is encapsulated in less than two pages of *The Scientific Image* (see van Fraassen 1980, 39–40). And he never subsequently provided a detailed defense of it. The importance of this type of explanation for the success of science, and the need to further explore its value, is evident from the fact that there is an extensive *critical* literature discussing it. It has become almost obligatory for any realist to give at least a passing assessment of this type of anti-realist explanation for the success of science. In fact, the realists' characterization of van Fraassen's selectionist explanation seems to have taken on a life of its own, just as we saw with Larry Laudan's Pessimistic Induction. Many critics, though, misunderstand what it is that such an explanation aims to explain or can explain. I believe that the selectionist explanation has far more promise than is generally recognized.

In defending the selectionist explanation for the success of science, I argue that, contrary to what the critics claim, the selectionist can explain why it is that we have successful theories, as well as why it is reasonable to expect theories that have been successful in the past to be successful in the future. I also argue that the plausibility of the realists' explanation for the success of science rests on an inaccurate understanding of the nature of *predictive success*. The predictive success of our best theories is a *relative* success. But the relative success of our best theories does not warrant an inference to their truth or likely truth. This is a point I have made repeatedly in the previous chapters.

Selection and Success in Science

I want to begin by distinguishing the sort of selectionist explanation van Fraassen develops from related types of explanations, specifically other evolutionary explanations, for the success of science. About a decade before van Fraassen developed his selectionist explanation for the predictive success of our best theories, Karl Popper (1971) likened scientific change to evolutionary change, arguing that theories are subjected to a process of selection like the process of natural selection. According to Popper, the testing of theories serves to weed out unfit alternatives. That is, when scientists design experiments in their efforts to test their theories, those theories that fail the tests are proven false and thus discarded. Just as natural selection eliminates weak variants of a species, Popper believes that testing in science eliminates weak theories. This process, he claims, is the means by which scientists get ever closer to the truth, even if they never know how far they are from the truth.

There is a significant difference between Popper's selectionist explanation and van Fraassen's selectionist explanation. Whereas van Fraassen presents his explanation for the success of science as an alternative to the realist's explanation, Popper intends his explanation to be compatible with realism. Popper's explanation is thus a *realist-selectionist explanation*. Van Fraassen's explanation, on the other hand, is an *anti-realist-selectionist explanation*. This will become significant later, as numerous realists insist that van Fraassen's selectionist explanation is also compatible with a realist explanation for the success of science.

Other philosophers have developed selectionist explanations for the success of science that do not focus on the theory as the unit of selection. David Hull (1988), for example, argues that the institutions of science are structured in such a manner as to ensure that scientists are able to realize their epistemic goals. Hull never specifies what exactly he takes the epistemic goals of science to be, and he studiously avoids engaging in the realism/anti-realism debate, a debate he regards as futile and sterile. Hull suggests that, given the reward structure in science, certain types of behaviors are encouraged, and these types of behaviors are, luckily, the same sorts of behaviors that tend to lead scientists to produce research results that are deemed to be useful to their peers. According to Hull, various types of fraudulent behavior, for example, cooking one's data, are quite uncommon, as those scientists who engage in such behavior will produce results that cannot be replicated (see Hull 2001). Other scientists will ignore the research produced by scientists they suspect of engaging in such behavior, for those research results could infect their own research. Hull is not so naïve as to think that scientists are perfect. But he does think that, given the way the institutions of science are structured, many forms of deviant behavior are discouraged, for such behavior will undermine a scientist's own scientific career.[1]

Common to Hull's and van Fraassen's explanations is a concern to explain the success of science by citing the operation of a selection mechanism. But as noted above, unlike van Fraassen, Hull does not treat the theory as the unit of selection. In principle, Hull's explanation, unlike Popper's, is compatible with van Fraassen's explanation for the success of science.

Another *seemingly* related project is the sort of evolutionary epistemology associated with Donald Campbell. Campbell (1977) sought to explain how *natural selection* has made humans prone to get at the truth, and this has given rise to an extensive research program involving both psychologists and epistemologists. The psychologists and epistemologists engaged in this research program seek to develop explanations of particular processes that have a genetic basis, for example, sight, in an effort to explain how it is that we are able to develop an understanding of the world. This project is quite different from van Fraassen's in two respects. First, unlike van Fraassen, these psychologists and epistemologists uncritically assume that our understanding of the world is an accurate representation of the underlying structure of the world. In this respect, their project is more compatible with scientific realism. Second, they invoke the operation of *natural* selection rather than some other selection mechanism that is merely similar to natural selection.

Van Fraassen does not suggest that we are *by nature* inclined to develop predictively accurate theories. Rather, his claim is that *despite*

---

[1] Robert K. Merton (1973), a sociologist of science, developed an account similar to Hull's, emphasizing the important role of the social structure of science in enabling scientists to realize their epistemic goals.

the fact that scientists may not be developing true theories, given the practices constitutive of scientific research, it is not surprising that they develop predictively accurate theories. According to van Fraassen, theories that are not accurate are unlikely to survive. Given a choice between a scientific theory that generates accurate predictions and another that does not, no scientist will waste her time working with the theory that generates inaccurate predictions. Thus, other things being equal, theories that generate inaccurate predictions will be discarded or replaced by better theories, theories that enable scientists to generate more accurate predictions. There is no need to invoke the truth of the theory in order to explain the outcome.

The Critics' Concerns

In this section, I present five criticisms that have been raised against van Fraassen's selectionist explanation. These are common criticisms. Most of them have been raised by a number of critics. The various criticisms draw attention to the challenges facing those who wish to defend an anti-realist selectionist explanation for the predictive success of our best theories. My aim is to address these criticisms in the remainder of the chapter. Let us now consider the criticisms.

First, some critics allege that

(1) the selectionist explanation for the success of science can only explain *past* successes and gives us no reason to think that theories that have been empirically successful in the past will *continue* to be successful in the future.

Simon Blackburn raises this concern. Given the selectionist explanation, Blackburn believes there is *no reason* to think that past success is an indicator of future success (2005, 178). Peter Lipton also raises this concern, arguing that, given the selectionist explanation for the success of science, "the real miracle is that theories we judge to be well supported go on to make successful predictions" (2004, 194). Stathis Psillos also raises this concern. As Psillos explains, "there is no warrant that [theories that have survived the selection process] will be successful in the future" (1999, 97). Without such warrant, the critics believe that van Fraassen's selectionist explanation is not much of an explanation at all.

The second concern is as follows:

(2) the selectionist explanation is compatible with a realist explanation for the success of science, so it is not a threat to the realist explanation.

In the previous section, I mentioned that Popper, in fact, thought that a selectionist explanation is compatible with realism. Others have made similar suggestions, but with the explicit intention of undermining van Fraassen's anti-realist explanation. Andre Kukla, for example, argues that "truth and evolution are not explanatory rivals" (1996a, S299). Alan Musgrave also raises this concern, arguing that "van Fraassen's Darwinian explanation ... can be accepted by realist and anti-realist alike" (1988, 242). Those who raise this concern claim that realists and selectionists are concerned with different issues. Selectionists merely aim to explain why our current theories are predictively successful. Realists, on the other hand, aim to determine either (i) what is common to all successful theories or (ii) why a particular theory is successful.

The third and fourth concerns are corollaries of the second. They are as follows:

(3) the selectionist explanation is not sufficiently deep, for it does not explain what is common to *all* empirically successful theories,

  and

(4) the selectionist explanation does not explain why *any particular* successful theory is successful.

These concerns are intended to compel us to see the superiority of the realists' explanation for the success of science. Whereas the third concern presumes that there is some common cause responsible for the success of our current best theories, the fourth concern makes no such presumption. The fourth concern draws attention to the fact that the selectionist explanation leaves unexplained why any particular theory is successful.

There are a number of philosophers who raise both the second concern and either the third or fourth concern. Lipton, for example, claims that "the truth explanation and the selection explanation are compatible" (2004, 193). Consequently, he insists that one *can* accept both (193). Moreover, he argues that van Fraassen's explanation "does not explain why a particular theory, which was selected for its observational success, has this feature" (194). As far as Lipton is concerned, until we know why each successful theory is successful, we really do not have an adequate understanding of the predictive success of our best theories.

Philip Kitcher also raises this concern. He notes that in biology, Darwinians aim to identify "the *genetic* characteristics that endow organisms with high Darwinian fitness" (1993, 156; emphasis added). Similarly, he believes that philosophers of science want to know "the *generic* characteristics that endow theories with great predictive and explanation power" (156; emphasis added). But Kitcher believes that van Fraassen says nothing on this important matter. Further, Kitcher suggests that an explanation for the success of science that cites the characteristics common to all successful theories need not conflict with van Fraassen's selectionist explanation. Psillos also believes that the selectionist explanation does not account for "the deeper common traits in virtue of which the selected theories are empirically successful" (1999, 96).

Similarly, Jarrett Leplin (1997) argues that to understand the success of science, it is not sufficient to explain why the theories we do have are successful. According to Leplin, this is all the selectionist explains. But he insists that an adequate explanation for the predictive success of our best theories also needs to explain why each of the successful theories we have developed is successful. As Leplin puts the point, "to explain why *particular theories*, those we happen to select, are successful, we must cite properties *of them* that have enabled them to satisfy our criteria" (1997, 9). Again, the critics are insisting that a generic account is required, an account that identifies the attributes that are common to *all* successful theories.

The fifth concern raises the question of whether selectionist explanations can explain *anything* about the success of science. It is as follows:

(5) the selectionist explanation for the success of science cannot account for the fact that we have *any* successful theories *at all*.

Blackburn raises this concern. As he explains, van Fraassen's selectionist explanation for the success of science "does not tell us why we are so clever or so well attuned to things that any theory at all gets through the Darwinian sieve" (Blackburn 2005, 179). The implication is that the selectionist explanation provides no insight into how we managed to develop successful theories in the first place. That is, the selectionist explanation says nothing about scientists' capabilities with respect to developing theories.

These criticisms, if fair, suggest that van Fraassen's explanation is deficient. I aim to show that the critics have misunderstood the nature of the sort of selectionist explanation for the success of science that van Fraassen developed.

Defending a Selectionist Explanation

When van Fraassen developed his selectionist explanation, his aims were quite narrow. He merely sought to explain why it is that our current best theories are predictively accurate (1980, 219, note 34). Given the criticisms outlined in the previous section, it seems clear that van Fraassen's critics expect something more from his explanation. My aim in the remainder of this chapter is to clarify the *scope* of the selectionist explanation and show that van Fraassen's selectionist explanation for the success of science is stronger than the critics recognize. In the next section, I will show that van Fraassen's selectionist explanation is, in fact, a genuine competitor with the realist explanation for the predictive success of our current best theories. That is, contrary to the critics' second concern, I argue that the two types of explanations are not compatible. In this section, I will show that an anti-realist selectionist explanation like the one developed by van Fraassen can either adequately address the remaining concerns or show why, in each case, the concern is ungrounded. I will frame the discussion around addressing the following questions, which are numbered to correspond to the concerns discussed in the previous section:

(5) Why do we have any successful theories at all?
(4) What features are responsible for the success of any particular theory?
(3) What is common to all theories that are predictively successful?
(1) Why should we expect that the theories that enabled us to make accurate predictions in the past are likely to enable us to make accurate predictions in the future?

I will address the concerns in the order listed above, starting with (5).

(5) Can an advocate of the selectionist explanation for the success of science explain why it is that we have *any* successful theories at all?

Yes, it is more or less guaranteed that the theories that scientists accept will be successful to some degree. Recall that predictive success is not a categorical quality. We saw this earlier, in Chapter 8. Predictive success or accuracy is relative in two respects. First, predictive success is relative to the accepted standards. Typically, as a scientific field develops, the standards become more demanding. Consequently, the predictive accuracy that would be deemed satisfactory early in the history of a scientific field may not be tolerated later. As we saw in Chapter 1, the accuracy expected of Copernicus' theory was less demanding than the accuracy expected in astronomy a century later, after Brahe, Kepler, and Galileo made their contributions to the field. But the accuracy expected of European astronomers working 150 years before Copernicus was even less demanding. Peter Dear claims that in the Middle Ages, "generally, the desideratum was to locate planets within the correct zodiacal sign, i.e. to within fifteen degrees" (2001, 171, note 2). That is a rather low threshold for predictive accuracy. Indeed, as mentioned earlier, by Copernicus' time, when astronomers made predictions of the locations of the planets, they were expected to be accurate to within 5 degrees (see Gingerich 1975b/1993, 196).

Not only do the standards of accuracy change over time, but scientists *construct* their own standards of success. Some caution is in order here. Each individual scientist is not at liberty to set her own standards. Rather, the research community sets the standards, and it is done with an eye to the degree of precision or accuracy scientists in the field have achieved so far. But given that scientists set their own standards of success, it should not surprise us that our current theories are successful. Even if there is no theory that meets scientists' expectations, that is, no theory that successfully passes through the selectionist sieve, scientists can always alter the sieve. That is, they can change the standards to ensure that at least some theory passes through. Granted, scientists will not always alter their standards when none of the available hypotheses meets their expectations. Sometimes they will reject all available hypotheses. But they are at liberty to alter their standards, and doing so is not inherently unscientific.

To some extent, it seems that the criterion of success is the result of social consensus in the research community. That is, the degree of accuracy that is deemed acceptable is determined by what one's fellow researchers accept as accurate. And such standards change with developments in methodology and instrumentation. Again, as we saw in

Chapter 1, Tycho Brahe's innovations in observational methods in astronomy in the late 1500s altered the standards for all astronomers thenceforth. An astronomer was not at liberty to ignore the new standards and continue to work with the earlier standards. Brahe changed the field for everyone. Developments or changes in methodology and instrumentation will have similar effects in all sciences. As Peter Galison has shown, twentieth-century particle physics was affected by such developments, as was early modern astronomy (for the case of modern physics, see Galison 1987). Similarly, a zoologist working in the twenty-first century cannot ignore data derived from DNA analyses when constructing taxonomic trees, even if she does not employ such methods herself.

As noted in the previous chapter, some realists have seized on the fact that our theories typically become increasingly more precise as a field develops. They claim that even if our theories are not wholly true, this pattern of development suggests that we are getting increasingly closer to the truth (see, for example, Kitcher 1993; Psillos 1999; Harker 2010). But this inference is not warranted. The increasing accuracy that characterizes the development of a field does not necessarily indicate that we are getting closer to the truth. To see why, let us consider the second respect in which scientific success is relative.

Success in science is a comparative notion. A predictively successful theory is, to some extent, just a theory that predicts better than the competitors. Given that evaluations of accuracy are comparative, we are hardly warranted in claiming that a theory is true based on the fact that it is predictively superior to the theory it replaces. Consider the following. In the late Middle Ages, the Ptolemaic theory of planetary motion was predictively superior to its main competitor, a theory employing homocentric spheres (see Duhem 1908/1969, chapter 4; see also Lattis 1994, 114). Ptolemy's theory was more accurate, to a large extent because it employed eccentric circles, equant points, epicycles, and deferent circles in its planetary models. But the fact that the Ptolemaic theory was predictively superior to its competitors hardly warranted the conclusion that it was true, or even closer to the truth than the competitors. As van Fraassen notes, when scientists choose the predictively superior theory, they may be choosing *the best of a bad lot*.

As we saw in Chapters 3 and 4, to infer that the best theory is true or approximately true, we erroneously assume that "we are by nature predisposed to hit on the right range of hypotheses" (van Fraassen 1989, 143). That is, it is to assume that we are choosing from a set of theories that contains the true theory. The most that we are warranted in inferring when we find that one theory is predictively superior to others is that the superior theory is more likely true than the alternatives it is compared with. But that does not imply that it is likely true. This point was argued for at length in Chapters 3 and 8.

In summary, contrary to what the critics suggest, the selectionist can offer insight into why we have successful theories. A successful theory is one that (i) meets scientists' expectations and (ii) is superior to the competitors it is compared with. Given that scientists' expectations are largely determined by the theories they have actually developed, success is almost guaranteed. Scientists will generally regard as successful the theory that is more accurate, other things being equal. Consequently, not surprisingly, such success does not support the belief that our theories are true with respect to what they say about unobservable entities and processes.

Let us now consider the next concern:

(4) An adequate explanation for the success of science must identify what features are responsible for the predictive success of each particular theory.

The critics are mistaken on this point. The selectionist who aims to explain the success of a particular theory is much like the biologist who aims to explain the persistence of a particular species. The biologist needs to identify the particular features of the species that enabled it to survive in the specific environment in which it survives, as well as features of the competitors it had to contend with. In the biological world, a species' competitors may include not only other species, but also other varieties of its own species, that is, subpopulations that have different suites of features that may give them different chances of survival. Similarly, the selectionist in the philosophy of science needs to identify the specific features of the environment in which the selected theory came to dominate. For example, the selectionist needs to identify the prevailing standards of accuracy and methodological practices and norms. As well, the selectionist needs to identify the features of the competitor theories.

It should be noted that features that explain the success of a theory at one time may be irrelevant to its success at another time. This is because, in the course of the history of a scientific field, theories will confront different expectations. Success, as we saw above, is a relative notion. This means that the success of a given theory at different times needs to be explained by different features. Different features of a particular theory become salient when compared with different competitors. Further, as some scientific problems become obsolete and new problems emerge as a field develops, the expectations of a theory are apt to change. Given that scientific fields are subject to constant change, the sort of generic explanation of the success of our theories that some realists desire might not be possible.

We have already provided a partial answer to the critics' next concern:

(3) An adequate explanation for the success of science needs to explain what is common to all predictively successful theories.

No. A generic explanation, an explanation that identifies features common to *all* predictively accurate theories, is not reasonable to demand. Just as we do not expect the persistence of different biological species to be a consequence of the same features, except in the most general terms, we should not expect the predictive success of all of our best theories to be a consequence of the same feature or features. Insofar as the selectionist will offer a generic account of the predictive success of our theories, it will be by citing the fact that these are the theories that have survived the selection process.

Further, as we saw in previous chapters, given the history of science, it is not reasonable to infer that all, or even most, predictively successful theories accurately represent the world. As noted earlier, scientists have frequently generated true predictions from false theories, even true predictions of novel phenomena. This should not surprise us, given what we learned above about the relative nature of success. What theories count as successful is determined, to a large extent, by what scientists are prepared to tolerate and what they have to choose from. Consequently, whatever all successful theories may have in common, it is not the fact that they are true. Thus, contrary to what some of van Fraassen's critics suggest, it is doubtful that the predictive success of our theories is a function of some feature or features common to *all* successful theories.

I am prepared to acknowledge that the realist critics are correct to insist on having more details about the mechanism responsible for the selection of our best theories in science. But indicating a need for further development is quite different from insisting that the explanation is bankrupt. Realists have exaggerated the problems with the selectionist explanation for the success of science.

Let us now consider the next concern of the critics:

(1) An adequate explanation for the success of science must explain why theories that have been predictively accurate in the past are apt to be predictively accurate in the future.

I can meet this demand. What the selectionist needs to show is that it is *reasonable* to expect accurate predictions in the future from the theories that have enabled us to make accurate predictions in the past. I believe that such a demand can be readily met, provided we take the comparison with natural selection seriously. In the biological world, it does not surprise us when a species that has survived until now continues to survive in the future. Given its past success, we are apt to be surprised if it *does not* continue to survive, unless we are aware of changes in the environment or the arrival of new competitors that would alter the species' prospects of survival. We think we are *warranted* in expecting the species to survive, even if we may not understand what traits or environmental factors contribute to its survival. Only someone who believes that the problem of induction, the problem of *proving* that the future will resemble the past, needs to be addressed first would think such an inference irrational.[2]

The situation is similar with respect to past successful theories. Though we may not know exactly why our past successful theories are successful, the fact that they have been successful inclines us to expect that they are apt to continue to be successful. Indeed, we are not *unwarranted* in having such an expectation, provided the theory is employed in ways similar to the ways it was employed in the past and there are no new competitors.

But theories that have enabled us to make accurate predictions in the past sometimes fail us. Similarly, in the biological world, species that

[2] It is worth noting that van Fraassen is skeptical about inductive inferences. But given his permissive notion of scientific rationality discussed earlier, it is not irrational for a scientist to believe anything that is consistent with her other beliefs. Thus, from the perspective of van Fraassen's standards, expecting that a particular species will survive given that it has survived in the past would not constitute an irrational expectation.

have long survived are sometimes driven to extinction. This fact, though, does not undermine our warrant in believing that, other things being equal, a particular species will persist, or a particular successful theory will continue to enable us to make accurate predictions.

Sometimes those who raise this criticism assume that the anti-realist, and van Fraassen in particular, is an instrumentalist of sorts. The underlying assumption is that instrumentalists are incapable of explaining the continued predictive success of our theories. Such critics grant that we need not be surprised that even an instrumental theory can enable us to make accurate predictions of many observable phenomena. After all, our theories are *designed* to account for bodies of data. They are made to fit. But the critics claim that it is surprising when a theory that does not purport to accurately represent the world is able to make accurate predictions of *novel phenomena* it was not designed to account for (see Musgrave 1988; Leplin 1997). These critics believe that such successes are not at all surprising for the realist. The truth of our theories explains their success. But these critics believe that the selectionist, insofar as he is an instrumentalist about theories, gives us no reason to expect our theories to continue to be successful.

I have two replies to this criticism. First, I am not interested in defending instrumentalism, and not all anti-realists are instrumentalists. Certainly, most contemporary anti-realists do not identify as instrumentalists. Van Fraassen makes it quite explicit that he is not an instrumentalist when he states that theoretical claims, that is, claims about unobservables, are candidates for being true or false. In contrast, instrumentalists claim that it is inappropriate to evaluate theories as true or false. Instead, they are to be evaluated as either useful or not, much as a tool is evaluated as effective or ineffective.

Second, as we saw in Chapter 9, contrary to what the critics suggest, scientists *have* generated accurate *novel* predictions from false theories, predictions that were subsequently vindicated (see Carrier 1991; Lyons 2002; Vickers 2013). Why false theories are able to do this is mysterious, as the critics suggest. But apparently it is not a miracle. The facts are as they are. Scientists can sometimes generate predictions of novel phenomena from false theories. Hence, as noted in the previous chapter, contrary to what the realists would like us to believe, novel predictions do not settle the case against the anti-realists.

Van Fraassen and I are not committed to the view that our theories are, in fact, false. But we do not think that the evidence supports the realists' conviction that our predictively successful theories are likely approximately true. And whether our theories are approximately true or not, we have some reason to think that successful theories will continue to be successful, at least in the short term.

Are the Selectionist and Realist Explanations Really Compatible?

As we saw earlier, some realists are willing to accept a selectionist explanation, but only one that is compatible with realism, like the one developed by Popper, for example. My aim in this section is to explain why the particular type of selectionist explanation that I have been defending, the type initially developed by van Fraassen, is a genuine competitor with the realists' selectionist explanation. Further, I aim to show that this anti-realist explanation is superior to the realist explanation. Hence, I will be comparing an anti-realist selectionist explanation and a realist non-selectionist explanation.

As the critics suggest, a selectionist explanation is, in principle, compatible with realism. But such an explanation is fundamentally different from the sort of selectionist explanation that anti-realists defend. According to the realist selectionist explanation, predictively accurate theories are selected because they are true or approximately true.[3] But according to the anti-realist selectionist explanation, predictively accurate theories are selected because they are superior to the competitors they are compared with. According to the anti-realist, such theories *could* be false with respect to what they say about unobservable entities and processes. Scientists, as van Fraassen notes, may be choosing from a bad lot. Given the fact that scientists are generally choosing from a small set of theories, and there is no compelling reason to think that the set includes the true theory (or even an approximately true theory), the inference from success to truth is unwarranted. This is why the anti-realist selectionist explanation for the success of science is not compatible with realism.

As is evident from the No Miracles Argument, discussed in the previous chapter, the realist aims to account for the success of only

---

[3] Popper would not make such a strong claim. Though he thinks that scientists *aim* for the truth, he believes that their method for getting there is to eliminate false theories through testing (see Wray 2015b). Testing, though, only enables scientists to determine which theories are false. But the realists who attack the anti-realist selectionist explanation want to draw the inference that our successful theories are likely true or at least approximately true.

those predictively successful theories that are true or approximately true. The realist, after all, wants to take success as a reliable indicator of the truth or approximate truth of a theory. The challenge the realist faces is in distinguishing true theories from false theories, or, more precisely, false theories that are thought to be true. The anti-realist selectionist explanation, on the other hand, aims to account for the predictive success of *all* predictively accurate theories, those predictive successes that are a consequence of being generated from theories that are true or approximately true and those that are generated from theories that are false (but perhaps thought to be true). Thus, some of the successes the anti-realist aims to explain will not be explained by the realist. Hence, it is misleading for realists to claim that the two types of explanations are compatible. They are, in fact, genuine competitors. Whereas anti-realists believe that the same feature or features may explain why both true and false theories are predictively successful, realists do not claim to be able to explain the predictive success of false theories. At any rate, they do not think the predictive success of false theories is to be explained in the same terms as the predictive success of true theories.

Having distinguished between a realist selectionist explanation and an anti-realist selectionist explanation, I want to argue that the anti-realist explanation is superior to the realist explanation in at least three respects. Significantly, these are considerations that realists should regard as valuable.

First, the anti-realist selectionist explanation has a *broader scope* than the realist selectionist explanation. The anti-realist explanation explains the predictive success of *all* predictively successful theories, past and present, true or false. On the other hand, the realist explanation for the success of science, the explanation that cites the fact that our predictively successful theories accurately reflect the unobservable structure and processes of the world, only aims to account for some successes, the successes of true theories.

Second, the anti-realist selectionist explanation explains the phenomena in a *parsimonious* manner, citing the operation of a single mechanism, a selection mechanism. The mechanism that figures in the anti-realist explanation does not discern between our predictively successful theories that are in fact true or approximately true and those that are false. It merely selects for predictively accurate theories, given the actual theories considered by scientists. On the other hand, the realist will attribute the predictive successes of true theories to their being true and the predictive successes of other theories to some other as-yet unspecified features.

Third, as noted in the previous chapter, advocates of a realist explanation must always be prepared to revise their explanation for any particular successful theory should it later be discovered that it does not accurately represent the unobservable structure and processes of the world. That is, the realist must retract his explanation of the success of theories that are successful today and *seem* to be true but are subsequently replaced by competitor theories, provided the new theories make radically different ontological assumptions than those made by the replaced theories. The anti-realist, on the other hand, is not prone to this problem. She readily acknowledges that even false theories can be successful, and her conception of success is compatible with predictively accurate but false theories. And even as science progresses and past successful theories are discarded and replaced by theories that make radically different ontological assumptions, the anti-realist need not retract her earlier explanation for the success of any particular past successful theory.

Given that an anti-realist selectionist explanation has these virtues, it seems that even the realist should recognize that such an explanation is superior to a realist selectionist explanation.

I want to briefly address a concern that I anticipate realists may have with my argument in this section. Given my criticisms of the theoretical values in Chapter 8, my appeal to scope and simplicity in my efforts to defend an anti-realist selectionist explanation may seem ironic, if not hypocritical. The irony, though, is only apparent. These traits, simplicity and breadth of scope, are the traits that *realists* regard as highly valuable when assessing theories. So I am arguing that by *the realists'* own standards, the anti-realist selectionist explanation is superior to their appeal to truth.

In summary, I have sought to correct some popular misunderstandings about the nature of the type of selectionist explanation for the success of science that van Fraassen presents in *The Scientific Image*. Though van Fraassen's remarks there are quite sketchy, realists have been discussing and critically evaluating this type of explanation with considerable enthusiasm. As realists have had the most to say about it, their characterization of the explanation has come to dominate the scholarly literature. This is unfortunate, for what has emerged is a distorted view of the nature of an anti-realist selectionist explanation for the success of science. I have aimed to present a more balanced picture of what an anti-realist selectionist explanation claims.

In this chapter, I have developed and defended a selectionist explanation that is incompatible with a realist explanation. My explanation is a genuine competitor to a realist explanation for the success of science. I have also shown that an anti-realist selectionist explanation for the success of science is more robust than realists have thought. The anti-realist selectionist can explain why it is that we have successful theories, as well as why it is reasonable to expect past successful theories to be successful in the future. Further, I have argued that successes in science are relative successes, relative to the currently accepted standards and the available alternatives. Thus, despite the growing enthusiasm for the latest popular versions of realism, Structural Realism and Selective Realism, realists cannot claim victory yet. The anti-realists' selectionist explanation is still a viable contender.

The anti-realist selectionist insists that the predictive success of our current best theories is a consequence of the fact that theories that do not save the phenomena, that is, theories that fail to account for what has been observed, tend to be discarded. No scientist can afford to waste her career working with such theories. A consequence of the effective operation of the selection mechanism operative in science is that our current best theories tend to be predictively accurate, at least by the lights of the accepted standards of accuracy. But the anti-realist selectionist also insists that this explanation, though compatible with extensive scientific progress with respect to our knowledge of observables, in no way implies progress in our understanding of the underlying unobservable entities and processes.

The selection mechanism operative in science, like natural selection in the biological world, is essentially an eliminative process, getting rid of the least fit alternatives. Thus, contrary to what realists seem to suggest, it will not necessarily drive scientists to select the true theory. In fact, unless we happen to be choosing between a set of theories that includes the true theory, the mechanism has no means to push us in that direction. Selection in science, like selection in the biological world, merely works with what it encounters. Its creative power is in ensuring that the best of what is available is not eliminated.

11 | *How Are False Theories Able to Make True Predictions?*

The fact that a theory generates successful predictions is the chief piece of evidence that realists cite in support of scientific realism. Many realists, including Alan Musgrave and Jarrett Leplin, emphasize the evidential import of predictions of novel phenomena (see, for example, Musgrave 1988; Leplin 1997). Some realists emphasize the fact that a theory can generate predictions on a routine basis over long spans of time, and some emphasize the precision of the predictions of our best theories. John Wright, for example, lists a number of very precise "empirical confirmations of [quantum electrodynamics]," including the values of (i) the magnetic movement of electrons, (ii) the Lamb shift for the hydrogen atom, and (iii) the positronium spectrum (see Wright 2013, 13–14). All of these facts are thought to be easily explained by the realist's conviction that our theories are likely true or approximately true (see, for example, Brown 1985; Wright 2013). But all of these facts, the realist argues, seem rather improbable if our theories are false. As we saw earlier, these considerations are the basis of the realists' No Miracles Argument discussed in Chapter 9. Many realists believe that the continued success of false theories would require either some sort of miracle or some sort of cosmic coincidence (see Smart 1963/2009; Putnam 1975).

It seems the anti-realist must provide an explanation for how false theories can generate true predictions, and even true predictions of novel phenomena. Importantly, as mentioned earlier, not all anti-realists insist that our theories are false. Indeed, as we saw in the previous chapter, one of the principal contemporary proponents of anti-realism, Bas van Fraassen, is agnostic on the truth or falsity of our best theories (see van Fraassen 1980). His concern is with the realists' inference from the success of our theories to their likely truth or approximate truth. But it seems that if anti-realists *can* explain how a false theory can generate true predictions, then they effectively undermine the realists' inference from the predictive success of a theory to the truth of the theory.

In this chapter, I want to consider the reasons why a false theory is able to generate true predictions. I will focus on two particular cases, Ptolemy's theory of planetary motion and the chemical theory associated with Mendeleev that classified chemical elements according to their atomic weight. In examining these cases, I will draw attention to a variety of reasons why a false theory can be successful. I argue that the same sorts of considerations seem to explain the success of other theories as well.

Before proceeding, it is worth noting that there is some evidence that, as a matter of fact, scientists are not persuaded by successful predictions to the extent that many realists suggest (see Brush 1994, 135; Scerri and Worrall 2001). If successful predictions played the role that realists imply they should play, effectively determining which theories scientists accept, then the predictive success of theories should decisively resolve disputes in science. But this is not what we find in the historical record. Stephen Brush has identified cases from the history of science where:

(i) a theory was accepted despite the fact that its predictions were false;

(ii) a theory was rejected despite the fact that its predictions were true;

(iii) a theory was accepted independent of the confirmation of novel predictions; and

(iv) the retrodictions generated from the theory were given as much credence in the acceptance of the theory as the novel predictions derived from the theory (see Brush 1994, § 4).

Indeed, many of the most spectacular predictive successes of theories were made only after the theories had long been widely accepted. But that need not speak against the (alleged) fact that the truth or approximate truth of our theories explains their success. Brush's chief concern is with the descriptive issue: Do scientists value predictions of novel phenomena to the extent that philosophers of science imply they ought to?

Scerri and Worrall (2001) focus specifically on the role of prediction in the acceptance of Mendeleev's Periodic Table of Elements. They found "little support for the standard story that . . . predictive successes were outstandingly important in the success of Mendeleev's scheme" (see Scerri and Worrall 2001, 407). In fact, they note that "accommodations played an equal role" (see Scerri and Worrall 2001, 407).

To some extent, both the normative issue, whether prediction should play a significant role in theory evaluation, and the descriptive issue, whether it does in fact play a significant role in theory evaluation, are tangential to my chief concern, which is to explain how a false theory could be successful and generate true predictions.

Why Was Ptolemy's Theory Successful?

Ptolemy's theory of planetary motion, specifically his planetary models, which were used to generate predictions of where the planets would be on future nights, was relatively successful. Indeed, as noted in Chapter 1, it was as successful and accurate as Copernicus' theory, which was developed more than a millennium after Ptolemy's theory (see Gingerich 1975b/1993, 195–196, figures 1 and 2). Copernicus' theory did not attract the attention of astronomers because of its predictive superiority. Ptolemy's theory, though, is false in a number of very significant ways. It assumes that the Earth is more or less at the center of the cosmos. It assumes that the planets, stars, and Sun complete an orbit around the Earth each day. Read literally, the planetary models suggest that each planet moves on an epicycle around a deferent circle that carries the planet around the Earth.[1] And it assumes that the stars are embedded in a sphere made of quintessence that completes a rotation around the Earth each day. In all of these respects, Ptolemy's theory misrepresents the structure and dynamics of the cosmos.

Given the predictive success of the theory and the fact that it does misrepresent the cosmos, it is worth examining why Ptolemy's models were so successful. In the remainder of this section, I aim to show that there are a number of features of both the research environment and the planetary models that explain the success of the models, despite the fact that they are not accurate representations of reality, given what our current best theory suggests.

First, Ptolemy's models were predictively successful, in part, because the accuracy expected of predictions was then relatively low, at least compared to today's standards. It is easy to make too much of this, so

---

[1] It is doubtful that Ptolemy believed in the reality of epicycles and deferent circles, but numerous astronomers who worked in the Ptolemaic tradition, including Christopher Clavius, did believe in their reality.

it is worth clarifying what I am claiming. Today, the accuracy with which we would be satisfied with the prediction of the location of a planet is far more precise than the accuracy expected in Ptolemy's day. One must remember, though, that astronomers did not achieve a higher degree of accuracy with the first well-developed heliocentric theory either. As Gingerich notes, Copernicus' theory was no more accurate than Ptolemy's. So the inaccuracy of Ptolemy's models was not (wholly) a function of the fact that he accepted a geocentric theory.

Now, this is probably a typical situation in any science. Generally, standards of accuracy increase over time. It is important to recognize, however, that Ptolemy's models were very successful by the standards of his time, and even by the standards of the sixteenth century, when Copernicus developed his theory. So one reason false theories can generate accurate predictions is because standards of accuracy are determined, to some extent, by what the available theories can deliver. It would be Whiggish for one to suggest that those earlier theories were inaccurate because they failed to measure up to today's standards, given that the scientists who worked with those theories were in fact impressed with the level of accuracy they were able to achieve. Indeed, Christopher Clavius, a contemporary of Tycho Brahe, was even led to infer that Ptolemy's theory and models were true by virtue of their predictive accuracy (see Clavius 1581, in Duhem 1908/1969, 94). That is, Clavius made just the sort of inference that realists urge us to make.

Second, Ptolemy's models were accurate because they were deliberately designed to make accurate predictions, with specific mechanisms, like eccentric circles and equant points, introduced to account for anomalies or irregularities. By employing such mechanisms in his models, Ptolemy was able to create the most accurate models to date. Indeed, Ptolemy's theory was the culmination of a model-building tradition in the ancient Greek world. And, as emphasized above, these were not surpassed in their accuracy until the seventeenth century, with the work of Johannes Kepler (see Gingerich 1971/1993, compare figures 5 and 6).

Again, we cannot impugn Ptolemy because he introduced ad hoc devices into his models with the intention of achieving a greater fit between his theory and the world. Copernicus, after all, did the same thing. Copernicus used a variety of mechanisms in his planetary models, including eccentric circles, epicycles, and deferent circles (see Kuhn 1957, 169–171). And these were carefully introduced into his models in order to achieve a better theory-to-world fit. That is, they were added ad hoc.

More generally, Thomas Kuhn argues that such a practice is one of the main tasks of normal science, the sort of science that most scientists are trained for and spend their research careers doing (Kuhn 1962/ 2012, chapter 3). So the fact that a theory is able to accurately predict phenomena because it is deliberately designed to do so cannot be a reason to discount its predictive successes. Ian Hacking provides an apt description of the working scientist's approach to experimental work that recognizes the central role that *accommodation* plays.

Research scientists have theoretical models, speculative conjectures couched in terms of those models; they also have views of a much more down-to-earth sort, about how [an] apparatus works and what you can do with it; how it can be designed, modified, adapted. Finally, there is that apparatus itself, equipment and instrumentation, some bought off the shelf, some carefully crafted and some jerry-built as inquiry demands it. Typically, the apparatus does not behave as expected. The world *resists*. Scientists who do not simply quit have to *accommodate* themselves to that resistance. They can do it numerous ways. Correct the major theory under investigation. Revise beliefs about how the apparatus works. Modify the apparatus itself. The end product is a robust fit between all the elements. (Hacking 1999, 71)

With all these possible strategies for achieving a fit, it should not surprise us that sometimes the theory that has been made to fit is false and misrepresents the world in fundamental ways. Moreover, given the prevalence of the practice of accommodating theories to data, one should hesitate before drawing an inference from the predictive accuracy of a theory to its likely truth or approximate truth.

Third, Ptolemy's models were predictively successful, or at least regarded as such, because certain discrepancies between the theory and the world were simply ignored. For example, if one were to read Ptolemy's model of the orbit of the Moon literally, one would expect the size of the Moon to change dramatically in the course of its revolutions around the Earth. In fact, according to Michael Hoskin, given the size of the Moon's epicycle in the Ptolemaic model, "the height of the Moon above the Earth varied between 33 and 64 Earth radii. This ought to have resulted in its apparent diameter varying by a factor of nearly two" (Hoskin 1997a, 44). But such variance was not detected, at least nowhere near the magnitude that the theory implied.

Little fuss was made about this fact, and it was certainly not regarded as grounds for rejecting the theory. The models were appreciated for what they could predict, and often their shortcomings or limitations were just overlooked.[2]

The same thing can be said about Copernicus' models. There are phenomena that one would expect to observe, given the Copernican theory and planetary models, that did not impugn the theory or models even though they were not observed. Theories are often quite partial in what aspects of the world they represent (see Chakravartty 2007, 190–192). Consequently, when a theory is described as predictively successful, we are often ignoring aspects that are, strictly speaking, misrepresentations of reality. Hence, it would be odd to infer that a theory is true or approximately true when we know it misrepresents the world in a number of respects.

Fourth, Ptolemy's models introduced features that inadvertently "accounted for" some underlying but unknown causes. For example, in Ptolemy's models of the superior planets – Mars, Jupiter, and Saturn – the line running from the deferent circle to the planet on the epicycle was always parallel to the line running from the Earth to the Sun (see Ptolemy 1952, Book X, § 6; see also Hoskin 1997a, 44) (see Figure 5).

This was merely a stipulation on Ptolemy's part, and it had the effect of enabling him to create relatively accurate models of the superior planets' motions. It had the effect of tying the planets' motions to the Sun. This is rather fortuitous, given that our current best theory tells us that these planets in fact orbit the Sun. Clearly, it was not the fact that the superior planets orbit the Sun that motivated Ptolemy to introduce this stipulation into his models. After all, he did not believe that the superior planets, or any other planets for that matter, orbited the Sun. But the effect of this stipulation was to produce models that were quite accurate given the prevailing standards.

This case study shows that false theories can make accurate predictions, and can do so on a regular basis and over long periods of time.

---

[2] In his recent book on models in science, Axel Gelfert notes that "what calls out for an explanation ... is the continued success of some – by representational standards: egregiously – false models, while many other ('truer') models fall out of favour" (Gelfert 2016, 81). Gelfert further notes that idealization and abstraction, which are common features in scientific models, render "most models literally false as representations of a specific target system" (81).

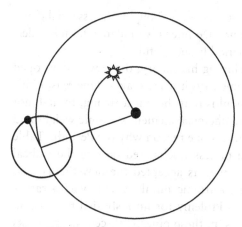

Figure 5 Ptolemy's model for the superior planets

As we see, there are a variety of reasons why this is so. Importantly, Ptolemy's theory was developed more than 1,500 years ago, and a more successful theory was not developed until the seventeenth century, with Kepler's significantly refined version of the Copernican theory. This should give pause to the realist who seems too hasty in drawing conclusions about the likely truth or approximate truth of our theories on the basis of their predictive success. Predictive success cannot be taken as unequivocal evidence in support of realism or the truth of a particular scientific theory. In other words, we have good reasons for doubting the realists' claim that if our theories are not true or approximately true, then it is a *miracle* that they can generate true predictions.

The Success of a False Chemical Theory

Other discarded theories were successful for the same sorts of reasons that Ptolemy's theory was successful. Consider, for example, the chemical theory that organized the various chemical elements according to atomic weight rather than atomic number, discussed at greater length in Chapter 7. Recall that before the discoveries of atomic numbers and isotopes in the early twentieth century, chemists identified chemical elements by their atomic weights. This proved to be very effective at advancing our knowledge of the chemical world, despite the fact that atomic weight is not as chemists generally thought, the essential defining feature of chemical elements. Despite this fundamental misunderstanding, the theory was remarkably successful.

First, like Ptolemy in developing his theory, those who developed this chemical theory made some fortuitous mistakes. One reason this discarded theory was successful is that there is a strong, though not perfect, correlation between a chemical element's atomic weight and its atomic number. In fact, that is one reason why the Periodic Table of Elements remained more or less unchanged despite the radical change that occurred when chemists accepted the new theory that ordered elements according to atomic number. This was a rather fortuitous accident. So just as Ptolemy fortuitously tied the motion of the superior planets to the sun, these nineteenth-century chemists fortuitously ordered the elements by a feature, atomic weight, that is strongly correlated with atomic number. But this same feature was responsible for some of the difficulties chemists encountered in the nineteenth century. For example, by regarding atomic weight as the defining feature of a chemical element, chemists made it challenging to come to terms with what came to be called isotopes, samples of the same element that differ with respect to atomic weight. Insofar as atomic weight defines an element, isotopes are an impossibility.

Second, as we saw earlier, some discrepancies between the theory and the world were simply ignored. In this case, a number of the suspected but hitherto undiscovered elements that Dimitri Mendeleev predicted on the basis of his theory were never discovered (see Scerri and Worrall 2001, 419 and 421). In fact, only half of the elements Mendeleev predicted were discovered (see Scerri 2011, 68, Table 21). These failed predictions, though, did not lead to the rejection of either Mendeleev's theory or his Periodic Table of Elements. Chemists, and scientists in general, tolerate a lot of discrepancies between theories and the world. So we should not be surprised that successful theories are later discovered to be false.

Third, some ad hoc adjustments were made that account for the success of Mendeleev's theory, which ordered the elements by their atomic weight. For example, Mendeleev and others took the liberty of reversing the order of some of the chemical elements on the Periodic Table, including the placement of tellurium and iodine. Such adjustments were contrary to the order of the elements determined by their atomic weight, but the decision was made on the basis of their other properties. Mendeleev even suggested that the atomic weights of these elements had not been properly determined. As Scerri explains, Mendeleev believed "that future experiments would eventually reveal an atomic weight ordering in conformity with his placement of tellurium before iodine" (see Scerri 2007, 125–126).

So, just as we saw in the case of the Ptolemaic theory, this chemical theory was quite successful, even though it was false.

In summary, I have identified four factors that explain why true predictions can be derived from false theories:

(1) Standards of accuracy change over time, and sometimes the standards are relaxed enough so that a false theory is able to meet the standards by which it is judged.

(2) Models and theories are intentionally designed to account for data, so it is not surprising that false models and theories are often quite successful.

(3) Often some of the shortcomings of a theory are disregarded, and thus apparent failures are not seen as failures.

(4) Finally, sometimes scientists introduce features into theories and models that fortuitously account for the effects of some feature that a successor theory determines to be causally relevant to the phenomena being modeled.

The analysis of the success of false theories that I present here should not be taken as a concession to or support for selective realism. Selective realists, like Stathis Psillos (1999) and Philip Kitcher (1993), urge us to distinguish between the working posits of a theory that are ultimately responsible for its success and the idle posits, which play no causal role in its success. They argue that what is retained in the successor theories through changes of theory are the working posits of earlier theories. In this way, they claim that even radical theory changes need not threaten an appropriately modest form of realism.

But as my analysis of the success of the Ptolemaic theory above suggests, it would be misleading to attribute its success to features that were retained in Copernicus' theory, the successor theory. Indeed, tying the motion of the superior planets to the Sun was "retained" in the Copernican theory, but it is only one of the features I identified as responsible for the success of the Ptolemaic theory. And as my analysis of the Chemical Revolution in the early twentieth century suggests, a number of the factors that account for the success of the earlier theory, the one that ordered the elements according to their atomic weight, were not retained in the new theory, which orders the elements according to their atomic number. Most importantly, the atomic weights of elements were no longer regarded as essential features.

Finally, it is worth noting that the argument developed here goes beyond the evidence presented by Martin Carrier, Timothy Lyons, and Peter Vickers, cited earlier. They have identified numerous false theories that generate true predictions, even vindicated predictions of novel phenomena. I have not focused on the distinction between novel and non-novel predictions. Rather, my concern has been with identifying general features that enable false theories to generate true predictions. I have identified some factors of false theories that explain the predictive success of these theories. What my analysis shows is that we do not need a miracle to explain why a false theory can generate true predictions.

12 | Discarded Theories
The Role of Changing Interests

Throughout the book, we have seen that looking back at the history of science, one sees that many successful theories were ultimately rejected, to be replaced by alternative theories, theories that made significantly different assumptions about the nature of the world than the assumptions made by the theories they replaced. I have discussed the following two example at length: (i) the replacement of Ptolemy's theory by Copernicus' theory and (ii) the replacement of the chemical theory that classified the elements according to their atomic weight by the theory that classified them according to their atomic number. It appears that revolutionary changes of theory like these are ubiquitous in the history of science (see Laudan 1981; but also Hesse 1976; Worrall 1989; Stanford 2001). In the 1950s, even Karl Popper noted that "in a time like ours ... theories come and go like the buses in Piccadilly" (Popper 1952/1963a, 125).

As I have stated, theory change presents scientific realists with a serious challenge. The realist wants to maintain that our best scientific theories are approximately true and that scientists are developing a picture of the world that is becoming increasingly more accurate. As we have seen, though, a change of theory often involves changes in basic assumptions about the world. For example, in Copernicus' theory, the Earth is no longer a unique body at the center of the universe, but a planet, like Mars, Mercury, and Jupiter. And with the new chemical theory that organized the elements according to their atomic number, two samples of the same element could have different atomic weights.

Reflection on the many once successful but since discarded theories in the history of science seems to suggest that today's best theories will meet a similar fate. If most past successful theories were discarded because they were false, it seems likely that some of the theories we accept today, though successful, will prove to be false in the future. Indeed, it seems like hubris on the part of scientists and philosophers of science to think that contemporary scientists have managed to do what their predecessors were unable to do, develop theories that are not only successful, but also approximately true with respect to the claims they make about unobservable entities and processes.

Some realists grant that many theories have been discarded in the past. But they insist that scientists have managed to develop better theories, *ones closer to the truth*, or *ones that better represent reality*. That is, each change of theory in a field is thought to be a step closer to the truth or to a more accurate representation of reality. Scientists have a fixed target set by nature, and each generation of scientists gets us closer to the target (see, for example, Bird 2000, chapter 6; also Kitcher 1993, 137). A theory is discarded when it is discovered that it missed the target. The realist thus offers a narrative of triumph, despite the history of science with its many discarded theories. This realist narrative of triumph is what I want to challenge in this chapter.

I want to offer some fresh insights into why the history of science is filled with discarded theories. I argue that the history of science is just as we should expect it to be, given the following two facts about science: (i) theories are always only partial representations of the world, and (ii) almost inevitably, scientists will be led to investigate phenomena that the accepted theories are not fit to account for. Together, these facts suggest that most scientific theories are apt to be discarded sometime and superseded by new theories that better serve scientists' *new* research interests. Consequently, it is reasonable to expect that many of the theories we currently accept, despite their many impressive successes, will be discarded sometime in the future.

I want to consider the extent to which we can explain the history of science without appealing to the realists' assumption that each new theory in a field typically is a step closer to the truth or a more accurate representation of reality than the theory that preceded it. I argue that new research interests will sometimes require radical changes of a kind that cannot be reconciled with most types of scientific realism. The realist can grant that changes of theory are apt to continue into the future. And some forms of theory change do not pose a threat to the realist's narrative of triumph. But what the realist cannot admit is that changes of theory introduce radical new conceptions of the world. Radical theory change is anathema to most forms of realism.

But I also argue that discarded theories are not aptly characterized as merely a sign of failure or a sign of some sort of shortcoming with science. Theories are often discarded because scientists are making advances in their pursuit of knowledge, solving research problems that their predecessors were unable to solve. In making such advances, scientists change their research interests, as the now-solved problem no longer needs their attention. Thus, discarded theories are often a sign of the proper functioning of science. Scientists are responding to their changing research interests.

Why Do Theories Fail?

In this section, I aim to reexamine the nature of scientific theories and their relationship to scientists' interests. It is worth emphasizing that I am not concerned here with either (i) social and political interests or (ii) the interests of individual scientists in advancing their careers. I recognize that such interests influence science and scientists, but these are not my concern here. Rather, my focus will be on the research interests that determine what sorts of issues a scientist investigates. No doubt research interests can be affected by those other sorts of interests, but I want to set aside consideration of those other types of interests.

Just to be clear, the following examples will illustrate the sorts of interests that will concern us. At one point in the history of astronomy, astronomers were concerned with the question of whether or not planets were self-illuminating. At one time this was regarded as an interesting and important scientific question (see Galilei 1612/2010, 93; Goldstein 1996, 4 and 7). This is, however, no longer a concern for astronomers. The question has been answered. Similarly, even in Johannes Kepler's day, a serious astronomer might have tackled the problem of explaining why there were only six planets. This is also no longer regarded as a genuine scientific problem, though it certainly was seen as one by Kepler. What I want to do is examine the effects that changes in these sorts of interests can have on scientific theories.

In outline, my argument is as follows: Every theory is only ever a partial representation of the world, thus every theory leads scientists to disregard some features of the world. Scientists' interests determine which features they disregard in their theories, and as they realize their research goals, their interests will change. Consequently, a theory that effectively served the interests of scientists at one time is apt to seem inadequate at some later time, when scientists have different research interests. At this later time, the theory is vulnerable to being discarded and replaced by a new theory that better serves current research interests.

I will begin by examining the nature of theories, for this provides the key to understanding why scientific theories have been discarded in the past and are apt to continue to be discarded indefinitely into the future. Much of what I will say here is neither new nor controversial. But I think that many philosophers have not thought through the implications of these claims.

Theories are partial representations of the world. They focus on and account for some features of the world but not others. The partial nature of theories is, in part, a consequence of the fact that theories often embody abstractions. When scientists introduce abstractions into their theories, they disregard aspects of the real world (see Chakravartty 2007, 221).[1] When scientists work with theories that embody abstractions, they are knowingly working with partial representations of the world. My concern here is not with the fact that such theories misrepresent the world (if in fact they do), but with the fact that they provide only a partial representation of the world. So my argument is not motivated by the concern that theories are false by virtue of being partial representations.

Philosophers of science have long been aware of the fact that theories are partial representations of the world. William James, for example, noted that "as the sciences have developed farther ... investigators have become accustomed to the notion that no theory is absolutely a transcript of reality, but that any one of them may from some point of view be useful" (James 1907/1949, 56–57). Similarly, Ernst Mach argued that "a theory ... always puts in the place of a fact something *different*, something more simple, which is qualified to

---

[1] Chakravartty (2007) provides a clear account of the difference between abstractions and idealizations. "An abstract theory is one that results when only some of the potentially many relevant factors present in a target system are taken into account" (Chakravartty 2007, 221). On the other hand, "an idealized theory is one that results when one or more factors is simplified ... so as to represent a system in a way it could not be" (221). The classic examples of idealized assumptions are frictionless planes in physics and rational agents in economics. Neither are to be found in the world. But as heuristics, they have proved indispensable. My focus will be on abstractions as they make our theories partial, accounting for some features of the world but not others.

represent it in some *certain* aspect, but for the very reason that it is different does not represent it in other aspects" (1892, 201).

Theories are partial representations because scientists are selective about what features of the world they attend to (see Mach 1892, 201; Poincaré 1913/2001, 182–185; Cartwright 1983; Giere 1988, 78–80; Longino 2001, for example). There is just too much information available to scientists, far more than they can effectively process. In order to make any sense of experience, scientists, and people in general for that matter, must be selective about what features they attend to (see Popper 1957/1963a, 61; Hempel 1966, 13). But unlike a layperson, a scientist is more deliberate and reflective about what features in the world she attends to. Whereas the layperson may often selectively attend to features uncritically, maybe even as a consequence of the evolutionary history of our species, the scientist consciously decides to take note of some variables and disregard others.[2] Clearly, guiding the scientist in her choices are theoretical assumptions about what the world is like. Carl Hempel vividly illustrates the selective nature of data collection in his account of Ignaz Semmelweis's attempt to determine why women in one ward of the Vienna General Hospital were prone to a higher death rate during childbirth than women in other wards. Each hypothesis Semmelweis considered led him to collect a different body of data and attend to different variables (see Hempel 1966, 3–6). Consider the data Semmelweis collected when he tested the hypothesis that medical students were bringing contaminants from the autopsies they conducted before their clinical work into the affected ward. Clearly Semmelweis was concerned with different data than the data he considered when he was testing the hypothesis that women were dying of fright from the presence of a priest attending to those who were dying in the hospital (see Hempel 1966, 4–5).

There are obvious advantages to working with theories that are partial representations of the world. By selectively attending to some features of the world and disregarding others, scientists are able to avoid being overwhelmed by information. This puts them in a better

---

[2] Popper discusses the evolutionary basis for the way animals divide their environments. A hungry animal discerns between food and non-food. An animal being pursued by a predator discerns between hiding places and escape routes (see Popper 1957/1963a, 61). Popper's examples are drawn from D. Katz's *Animals and Men*. Clearly, the layperson is more like an animal than a scientist in this respect.

position to detect patterns that might otherwise be difficult to detect. Further, by employing abstractions, scientists can work with theories that are more tractable. For example, calculating the positions of planets is made far simpler by disregarding the effects of the gravitational attraction of neighboring planets and by "treating the planets as point masses or homogenous spheres" (see Chalmers 2013, 223). I am not claiming that the only reason scientists introduce abstractions is to make a problem more tractable, but it is clearly one important reason.

Indeed, it is because scientists work with theories that are merely partial representations that they are so effective at realizing their research goals. This is a key point in Thomas Kuhn's *The Structure of Scientific Revolutions* (1962/2012; see especially chapters 3 and 4). Though paradigm-guided research leads scientists to be myopic, in periods of normal science this paradigm-induced myopia is generally an epistemic asset, focusing their attention on only those features of the world that really matter. In fact, Kuhn recognizes that this myopia is both an epistemic asset and an epistemic impediment. It is part of the "essential tension" that characterizes science. When scientists are working in a well-functioning research tradition, uncritically working with the accepted theory, paradigm-induced myopia can help them realize their research goals. They are determined to make nature fit into the conceptual boxes supplied by the accepted theory. But when persistent anomalies become intractable and a new theory is needed, paradigm-induced myopia can be a serious impediment to scientific progress. It can prevent scientists from seeing things that are relevant to accounting for the otherwise intractable anomalies.

There is a second, related feature of theories that contributes to their partial nature. Scientific inquiry is interest-driven. Which specific features a theory is designed to account for, as well as which specific features it disregards or brackets, is determined by the research interests of scientists. Theories are developed with specific research problems and goals in mind. And the research problems that concern a scientist will determine which features she takes account of and which ones she disregards.

For example, in early modern Europe, astronomers sought to account for the motions of the planets and stars as observed from the Earth. In developing their models, astronomers sought to account for particular features, like the direction of a planet's motion, including its periodic stations and retrograde motion, the period of a planet's cycle through the fixed stars, and the relative brightness of a planet in the course of its orbit. They focused on these features and disregarded others. For example, they made no attempt to account for the apparent color of the planets or the mass of planets and stars. These features were deemed irrelevant to their research goals. The mass of planets was regarded as irrelevant, to a large extent, because astronomers, unlike natural philosophers, were not concerned with causes. Nor did they suspect that a planet's mass was relevant to modeling its motion. Rather, their interests and efforts were primarily directed toward developing planetary models that enabled them to predict the locations of planets and the occurrence of conjunctions, eclipses, and other such phenomena (see Westman 1975; Duhem 1908/1969).

Given a different set of research interests, scientists would be led to account for different features than those they accounted for. For example, when Newton sought to develop a physical theory that unified terrestrial and celestial mechanics, the mass of celestial bodies became a relevant feature in his planetary models. His research interests, being different from those of his predecessors, dictated a change in the sorts of things he sought to account for.

Before moving on, I want to underscore my main point. Theories are limited in what they represent. Their limitations are not wholly or even principally liabilities. Rather, their limitations are what make them valuable. The partial nature of theories, however, can become an impediment when research interests change, and this provides a key to understanding why the history of science is a history of once successful but now discarded theories.

We need to resist the temptation to think that scientists only choose to work with theories embodying abstractions because they do not yet have the true theories. Scientists are not compromising when they make such choices. Abstractions play an indispensable and constructive role in science. They make doing the job of science tractable. Abstractions aid scientists in realizing their research goals. It is not profitable or insightful to think of a future science where abstractions will play no role in theorizing, as there is little evidence that such a future will ever exist.

The fact that a scientific theory is a partial representation of the world need not have a negative impact on the course of scientific research. Any specific abstraction that is built into a theory may never pose a problem for scientists. After all, the scientists who work with the theory may only apply it to phenomena that are largely unaffected by the features of the world that are not accounted for by the theory. But sometimes scientists find themselves studying phenomena that will be misrepresented by a theory, phenomena that are not easily accounted for given the conceptual resources supplied by the accepted theory. In these cases, the fact that a theory is only a partial representation may become a concern. Generally, however, in such situations a scientist's first impulse is not to discard a long-accepted theory. Doing so is costly. And generally scientists will want an alternative theory to replace the theory they are discarding. Most research cannot be conducted effectively without the aid of some theory or other. Whatever else theories might be, they are aids to research. Despite Kuhn's fame for emphasizing the role of theory change in science, even he recognized that a scientist's first impulse is to find a way to solve research problems using the resources of long-accepted theories. As Kuhn notes, "retooling is an extravagance to be reserved for the occasion that demands it" (Kuhn 1962/2012, 76). The decision to discard a long-accepted theory is not to be taken lightly. Other things being equal, there are strong incentives to continue to work with a long accepted theory.

Indeed, this conservative strategy often pays off. Often scientists will be able to salvage the long-accepted theory, accounting for anomalous phenomena by adjusting various parameters in their models. A modified version of the accepted theory may thus take the place of the older theory. Continuity may be restored. But sometimes it is not possible for scientists to merely augment the accepted theory to accommodate a new discovery. The long-accepted theory will prove to be a significant impediment to advancing scientists' research goals. In such cases, scientists may be led to discard the theory they have been working with.

Even in the normal course of research, scientists will sometimes consider discarding a long-accepted theory. It is worth examining how this sort of situation arises. The typical scientist's career is spent applying or extending an accepted theory (see Kuhn 1962/2012, chapters 3 and 4). Typically, scientists work with the conceptual resources supplied by a theory in an attempt to solve hitherto unsolved research problems, problems that are in a sense suggested by the accepted theory. This is the sort of work that science education trains scientists for (see Kuhn 1962/2012, 47). The working assumption is that the accepted theory is adequate to the tasks at hand. The challenging part of research is figuring out how the accepted theory can be applied to the specific phenomena one encounters (see Kuhn 1962/2012, 36).

In the pursuit of this goal, in their efforts to fit nature into the conceptual boxes supplied by the theory, two sorts of problems are apt to arise that may ultimately lead scientists to consider abandoning a long-accepted theory. First, they may encounter hitherto undetected phenomena that seem irreconcilable with the theory. X-rays, Neandertal remains, and novas raised challenges for long-accepted theories when they were first discovered. Indeed, we are still making striking discoveries about Neandertals and their relationship to our own species as scientists analyze and compare the DNA of Neandertal remains and of modern humans. For example, we now have evidence that the early ancestors of Asians and Europeans, but not Africans, interbred with Neandertals. This discovery significantly changes our understanding of our own species and its relationship to Neandertals and other hominids (see Green et al. 2010; Sankararaman et al. 2012; Vernot and Akey 2014).

Second, in the course of conducting research, after solving a series of research problems, a research community may inadvertently be led to raise new research questions that were unthinkable earlier. In these sorts of cases, the research community will have inadvertently developed new research interests and may find itself confronting research problems that cannot be adequately addressed with the conceptual resources of the accepted theory. In such a situation, the research community may be led to discard the long-accepted theory. For example, it was in the course of *extending* classical mechanics in an effort to solve a hitherto unsolved problem that Max Planck inadvertently contributed to the downfall of the accepted theory (see Kuhn 1987/2000, 25–28). What to Planck was intended as an expedient way to model black-body radiation ultimately led other physicists to discover problems with Newtonian mechanics (Kuhn 1987/2000, 27). In his work on the black-body problem, Planck assumed, merely for the purposes of his research, that radiation was not distributed continuously. Planck's assumption was motivated by a desire to make his research problem mathematically tractable. But as he worked on the black-body problem, he believed that radiation could in fact be distributed continuously (see Kuhn

1987/2000, 27). Inadvertently, Planck's research led to the downfall of classical mechanics.

Caution is in order here. I am not claiming that the research interests of scientists change completely from one theory to its successor. Despite his reputation, not even Kuhn believed this (see Kuhn 1992/2000, 113). My point is that, provided there is some significant shift in research interests in a research community, a theory that seemed adequate at one time may come to seem unacceptable later. As scientists direct their attention to different research problems, they change their research interests. The view I am defending here does not preclude extensive continuity in research interests through changes of theory. But even against a backdrop of extensive continuity, a long-accepted theory may prove inadequate for the new problems that come to concern the scientists working in a field.

In summary, what I have been arguing here is that, in the course of pursuing their evolving research interests, scientists may be led to discover the inadequacies of a long-accepted and hitherto empirically successful theory. Sometimes the critical evaluation of the long-accepted theory is related to the specific abstractions that figure in the theory. The features of the world that were disregarded earlier are now regarded as salient and cannot be disregarded any longer, given the research interests of the scientists working in the field. The change in status of the theory from harmless, even effective, partial representation to an impediment to research is sometimes a consequence of new emerging research interests.

Let me briefly address a criticism I anticipate. Earlier, I suggested that the replacement of a long-accepted theory by a new theory is not to be regarded as a failure. But above I described the replaced theory as inadequate. These claims are not inconsistent. What makes the old theory inadequate on the account I am presenting here is that researchers' interests have changed. So the shortcomings of the rejected theory are a function of the fact that scientists have changed the standards by which they are evaluating theories. Such changes of standards are inevitable as scientists change their research interests. Not surprisingly, often the scientists who experience a radical change of theory are apt to regard the discarded theory as a failure. By the new standards, standards that have evolved with the new alternative theory that takes the place of the discarded theory, the old theory is in fact deficient.

Whose Interests?

So far in my account of discarded theories, I have referred to changing research interests as if it were obvious *whose* interests I have in mind. It is worth clarifying whose changing interests are responsible for the many discarded theories in the history of science, as philosophers often speak of the interests of individual scientists, the interests of research teams, and the interests of scientific fields or specialties.

Elsewhere I have argued that scientific specialty communities are not agents. I have in mind here groups like endocrinologists, herpetologists, and inorganic chemists. These groups are not capable of having beliefs, intentions, or interests (see Wray 2007). More precisely, a scientific specialty does not have interests that are irreducibly the interests of the specialty. Consequently, we cannot expect a scientific specialty, taken as an irreducible whole, to change its interests. Individual scientists, though, do have interests. They choose to conduct research on one topic rather another. Research teams, groups of scientists who pursue research projects collaboratively, are also aptly described as having interests. Research teams may involve two scientists in either the same or different fields, or larger groups of scientists ranging from three to several hundred in number. Research teams, no matter what their size, must make choices about what research problems they will address. As a consequence, research teams are aptly described as having interests.

An individual scientist or a research team, however, cannot cause a theory to be discarded merely by changing research interests, nor even by deciding to no longer work with the long-accepted theory.[3] When a theory is discarded in the sense relevant to our concerns here, it is no longer accepted in the research community as a whole.

Some of Kuhn's critics raised the concern that he could not explain how a theory comes to be discarded by a research community. These critics claimed that, given Kuhn's account of scientific change, the accepted theory seems to have such a grip on the scientists working with it that it becomes inconceivable how the community could ever break free from it (see, for example, Laudan 1984a, chapter 4; also

---

[3] Perhaps the exception here is those enormous research teams that employ most of the scientists working in a field. When such a research team changes its interests, the field as a whole changes its interests. The field and the team are co-extensive. This, though, is probably rare, and may only happen in certain areas of physics.

Fisch 2017). According to these critics, Kuhn seems to suggest that scientists are incapable of seeing outside the accepted paradigm. This is an uncharitable reading of Kuhn's view, and it misrepresents the way scientists' interests change.

Typically, research interests change in a research community when hitherto outstanding problems are solved to the satisfaction of the members of the community. Then researchers turn their attention to other problems. In doing so, the research community changes its interests. The change of interests in a research specialty, though, is not a coordinated affair. That is, research communities do not operate by consensus conferences.[4] They do not convene meetings to determine what research problems they will address next. Rather, individual scientists and research teams will be compelled to address different research questions, and thus will be moved by different interests. Somehow, though, research communities generally manage to stay relatively focused on a circumscribed set of problems such that the community as a whole retains its cohesiveness and identity, even through episodes of theory change.

There are two factors that tend to ensure that a research community will persist through a change of theory, and that a new consensus will generally emerge after a long-accepted theory is discarded. First, the cohesion of a research community through a change of theory is secured, in part, by the fact that an individual scientist's own interests are affected and constrained by the interests of her peers and colleagues. If an individual scientist addresses problems that do not engage her colleagues, she will find that her research is ignored. David Hull has already highlighted this aspect of science. Hull argues that science is structured such that generally, the individual scientist's interests line up with the interests of science, the institution (see Hull 2001, chapter 5). Hull was concerned with explaining why scientists tend to uphold scientific standards and generally resist fudging data and engaging in other deceptive practices detrimental to science. The tendency for scientists to pursue research projects that are of interest to their peers is just one more manifestation of this happy coincidence that Hull identifies. A scientist may find herself without an audience for her research if her research interests depart too

---

[4] Consensus conferences are sometimes used in the medical sciences, but their purpose is to ensure that practitioners are aware of the consensus among researchers of the effectiveness of a new medical treatment option or therapy (see Thagard 1999, chapter 12).

far from the interests of the rest of the research community. The individual scientists and research teams working in a particular field are thus constrained in choosing what to investigate. Their own research interests must, to some extent, align and engage with the interests of the rest of the research community.

The second factor that explains why a research specialty is unlikely to fragment and a new consensus is likely to emerge after a theory is discarded is that scientists working in a specialty never face a boundless number of choices when consensus breaks down and the field is in crisis. Rather, there are often only two competing theories to choose from when they are looking for a new theory to guide them in their research. The Copernican Revolution in astronomy is exceptional in this respect. There were three well-developed alternative theories competing for the allegiance of European astronomers around 1600: the Copernican theory, the late Renaissance version of the Ptolemaic theory, and Tycho Brahe's theory. There were, in addition, other competitors, including a version of Brahe's theory that included the Earth rotating on its axis daily and the so-called Egyptian theory, which was Earth-centered, with Mercury and Venus, but not the other planets, orbiting the Sun as the Sun orbits the Earth (see Duhem 1908/1969, 83). Frequently, though, scientists are faced with a choice between only two competing theories. For example, in the early twentieth century, when chemists came to accept the theory that organized the elements according to their atomic number, they were only choosing between that theory and a single alternative, the theory that organized the elements according to their atomic weight.

Given that normally there are only two or three viable competitors to choose from, the fact that a new consensus emerges is no great mystery. As the competing theories are developed and revised in light of criticism from proponents of the competing theories, one theory is likely to emerge as the superior theory (see Wray 2011, chapter 9). Then it is no mystery at all why most of the scientists working in the field adopt that theory. Once a particular theory is adopted by most scientists working in a specialty, scientists' interests will be constrained and shaped by the accepted theory. They will, as Kuhn notes, tend to address research problems that the accepted theory is especially well suited to address (see Kuhn 1962/2012, chapter III).

Truth and Interests

In this section, I want to briefly address a criticism that I anticipate. I can imagine a determined realist suggesting that nothing I have argued for is incompatible with the realists' conception of the aim of science as the discovery of theoretical truth. Even when scientists' interests change, they are still concerned with getting at the truth. So the failure of a theory to measure up to the truth is what really explains why it was discarded, the realist critic claims.

Caution is in order here. In the literature on the realism/anti-realism debate, there is a lot of ambiguity in discussions of the aims of science (see, for example, the exchange between Rosen 1994 and van Fraassen 1994; and Rowbottom 2014). Bas van Fraassen rightly notes that the aims of *science* and the aims of *scientists* need not be the same (see van Fraassen 1994, 181). He claims that the aims of science are concerned with the criterion of success in science, whereas the aims of scientists are related to individual scientists' motives (see van Fraassen 1994, 182). Van Fraassen insists that the aims of scientists are irrelevant to understanding what divides realists and anti-realists. Darrell Rowbottom (2014) argues that the ambiguities surrounding the notion of "the aims of science" have created so much confusion in the realism/anti-realism debate that it is best to avoid any discussion of the aims of science. I think it is wise to heed Rowbottom's advice here.

What I have sought to show is that we can explain the fact that scientists have been led to discard theories that were long regarded as successful without invoking the notion of truth. On the account of discarded theories I have presented here, scientists are not always discarding theories because they discover that their theories are false. Rather, sometimes they are led to discard a theory when their research interests shift to such an extent that the long-accepted theory proves to be inadequate. The inadequacy of the long-accepted theory is a function of the fact that it no longer serves the current research interests of the research community. Hence, the salient feature is not that the theory discarded has now been discovered to be false. In fact, I suggested above that scientists are often knowingly working with theories that are only partial representations of the world. A partial representation need not be false. But by virtue of its being partial, it is bound to not account for some variables that may later prove to be of some consequence.[5]

It seems that part of what realists and anti-realists are disagreeing about is the relationship between truth and interests. Given that scientists cannot possibly expect to pursue all truths, some realists seem to think that scientists' interests merely serve to select which truths they pursue. Philip Kitcher holds such a view (see Kitcher 1993, 94). Realists such as Kitcher seem to think that the choice to not pursue other truths is inconsequential from an epistemic point of view. I believe that interests play a more pronounced role in science, and the choices scientists make to not account for some variables in their theories can be quite consequential, at least in the long run. Scientists' interests determine which truths they seek, and thus which variables they account for in their theories and models. Scientists' interests also determine which truths and variables can be disregarded. When scientists introduce abstractions, they do so because the abstractions serve their epistemic interests, focusing their attention on only those qualities that matter, given their current research interests.[6] And clearly, scientists' current assumptions about reality will affect which qualities they choose to attend to. This account of the role of interests in scientific inquiry fits the actual practice of science better than the account offered by the realist.

It is worth repeating a point I made earlier. I am not claiming that non-scientific interests or broader social interests do not shape scientists' research interests. No doubt they do. But these broader interests can only direct scientists so much. Scientists still need to determine what variables they will study or account for, and these choices will

---

[5] One might think that I am presenting a false dilemma here by suggesting that a theory is discarded because either (i) as is typically suggested, it is discovered to be false or (ii) as I suggest, it no longer serves the interests of scientists. This is not so. Scientists might discover that a theory both is false and no longer serves their interests.

[6] There are affinities between the view I present here and Rudolf Carnap's view in "Empiricism, Semantics, and Ontology." Carnap claims that the choice of a language or theory is a pragmatic choice. "The acceptance [of a language or theory] cannot be judged as being true or false because it is not an assertion. It can only be judged as being more or less expedient, fruitful, conducive to the aims for which the language is intended" (Carnap 1950, 31). Note the central role that he attributes to the aims of the people adopting the language or theory.

most often be affected by their conjectures about what sorts of factors are causally relevant to understanding the phenomena they are studying. In this respect, Kuhn was correct to claim that, in mature fields, scientists are shielded from the influence of broader social factors. The audience for scientists' research is, first and foremost, other scientists in their specialty, that is, their peers (Kuhn 1962/2012, 163). Indeed, some of the variables scientists work with may be determined by funding agencies. For example, the National Institutes of Health (NIH) in the United States might fund a grant program for research on diabetes among African Americans. Clearly, this puts some constraints on the variables that *need* to be accounted for. But there is much more that needs to be determined in designing a study that addresses this population, and this is left to the discretion of the scientists.

Unconceived Alternatives and Interests

The account of discarded theories I have presented here offers some new insights into an issue Kyle Stanford (2001; 2006) has drawn attention to, the existence of unconceived alternative theories, discussed in Chapter 5. Stanford's New Pessimistic Induction has proved to be one of the anti-realists' strongest arguments.

Recall that Stanford argues that "the history of scientific inquiry offers a straightforward inductive rationale for thinking that there typically *are* alternatives to our best theories equally well-confirmed by the evidence" (Stanford 2001, S9). Stanford gives a number of examples from a variety of scientific fields, including:

[1] the historical progression from Aristotelian to Cartesian to Newtonian to contemporary mechanical theories ... [2] the historical progression from elemental to early corpuscularian chemistry to Stahl's phlogiston theory to Lavoisier's oxygen chemistry to Daltonian atomic and contemporary physical chemistry ... [and [3] the historical progression] from Hippocrates's pangenesis to Darwin's blending theory of inheritance ... to Weismann's germ-plasm theory and Mendelian and contemporary molecular genetics. (see Stanford 2001, S9; numerals added)

Stanford emphasizes that "the evidence available at the time each earlier theory was accepted offered equally strong support to each of the (then-unimagined) later alternatives" (Stanford 2001, S9).

The more recently developed theories in each of these series of theories were not adopted *earlier*, because they were then unconceived. Stanford argues that reflection on the history of science, specifically the existence of unconceived alternatives, suggests that even today's best theories are likely to be replaced in the future by as yet unconceived alternative theories.

My argument above suggests that many of the various then-unconceived theories in the history of science are likely addressing different research problems than the problems addressed by the theories they replaced. Thus, part of the reason seventeenth-century natural philosophers abandoned Aristotelian physics is because they were developing research interests that were no part of Aristotle's concerns, and thus not fit to be accounted for by Aristotle's theory, even as developed in the Medieval period and the Renaissance. The physicists working in the mechanistic tradition associated with Descartes and Galileo, for example, wanted a physical theory that would offer insight into a number of phenomena that were either unknown to Renaissance Aristotelians or inadequately accounted for by the version of the Aristotelian theory accepted in the Renaissance, including (i) magnetism, (ii) how the planets stay in their orbits, and (iii) Harvey's discoveries about the physiology of blood flow, to name just a few. Consider phenomenon (ii). Until the late 1500s, the planets were thought to be embedded in spheres made of ether, so there was no need to explain how they stayed in their orbits. As we saw in Chapter 1, after careful observations of comets made in the 1570s and 1580s, the existence of such spheres was called into question, for comets appeared to cut through the (alleged) spheres (see Gingerich 2004, 155). Henceforth, there was a need to address, and interest in addressing, this new scientific problem. That is, once the existence of the spheres was called into question, explaining how the planets stay in their orbits became a scientific problem.

Similarly, Newton's concerns and interests were not the same as Descartes' concerns and interests. Each theory in the succession of theories in a scientific discipline addresses a different set of problems. Obviously, there is bound to be significant overlap and continuity. But more recently developed theories are developed by scientists concerned with different problems, some of which cannot be adequately addressed with the resources of the theory that is being abandoned.

Thus, it seems that unconceived alternatives are often not conceived earlier not because of a lack of imagination or creativity on the part of scientists, but because earlier scientists had different research interests. Obviously this is only part of the story. Research interests are also shaped by developments in instrumentation. For example, with the creation of the air pump, natural philosophers were able to investigate phenomena that were unimaginable to late-Renaissance Aristotelians. Seventeenth-century natural philosophers were able to examine the effects of the deprivation of air on various creatures, lit candles, and barometers (yet another new instrument). What Aristotle and the seventeenth-century Aristotelians had to say about air provided little or no insight on such topics. Not surprisingly, scientists turned to a new theoretical framework for insight.

My aim in this chapter has been to reexamine the history of science and reassess the significance of the pattern of theory change that seems to suggest that theories are apt to continue to be discarded indefinitely into the future. I have argued that the pattern of theory change that the standard Pessimistic Induction draws attention to is a natural consequence of the development of theories. As scientists develop their theories, they are led to ask research questions and model phenomena that their theories were not designed to answer or model. Rather than seeing the development of science as a march ever closer to the truth, it is more fruitful to see that scientists are constantly, though gradually, altering their research interests and agendas. And changes in research interests can lead scientists to evaluate theories that they once regarded as successful as inadequate. This is a significant factor in understanding why the history of science is a history of discarded theories.

13 | A Synthesis

Interestingly, there now seems to be substantial agreement among many realists and anti-realists about a number of the relevant facts about science and scientific change. Many realists acknowledge that there have been many instances of radical theory change in the past, where a theory that makes a particular set of ontological assumptions about various unobservable entities postulated to explain the observables is replaced by another theory that makes significantly different ontological assumptions. Many realists are even willing to acknowledge that such changes may in fact happen in the future. That is, even some of the successful theories we accept today will likely be replaced by alternative theories that make significantly different ontological assumptions than the assumptions our currently accepted theories make. The history of science is, to a significant extent, a history that involves numerous instances of radical theory change.

Contemporary realists and anti-realists also agree that the various sciences are very successful. In fact, both realists and anti-realists grant that our knowledge of the phenomena has grown markedly throughout the history of modern science, say, since 1600. The precision with which scientists can make predictions has gotten increasingly more exact with respect to many phenomena. Thus, the question of whether or not there has been progress in science is not at issue in the debate. In other words, those involved in the contemporary realism/anti-realism debate are not concerned with the sort of radical skepticism that much of the earlier literature on underdetermination was preoccupied with, a sort of underdetermination that seems to have originated with W. V. Quine. Contemporary anti-realists are not interested in trying to show that it is *logically possible* that our current best theories are radically mistaken. They are not worried about the possibility of human knowledge in general, or about our connection or lack thereof to the world. Moreover, realists rightly do not think it is worth their efforts to address such far-reaching skeptical worries. The skepticism that concerns those involved in the contemporary debate is a skepticism about theoretical knowledge only. It is a skepticism about our alleged knowledge of the unobservable entities that are posited to account for the phenomena.

Given the fact that there is such extensive agreement between realists and anti-realists about science, one might wonder why the realism/anti-realism debate has not passed and why a reconciliation has not yet been achieved. Indeed, there has been significant movement from more extreme positions on both sides to quite moderate realisms and anti-realisms in recent decades. Fortunately, these concessions have not been bought at the cost of a new scholasticism, that is, by making finer distinctions that really add nothing substantive to the discussion.

But there are good reasons for the debate to persist. These facts, as clear as they are to both sides in the debate, seem to admit of realist and anti-realist interpretations. On one side are the realists, who see significant progress and continuity in science despite the history of discarded theories with their mistaken ontologies. The continuity that is evident through theory change, realists claim, is evidence that scientists are getting at some sort of truth with their theories.

On the other side are the anti-realists, who acknowledge the empirical successes of science, even the growth in our knowledge of the phenomena, but deny that the growth of knowledge is connected in any sort of significant, systematic way with changes of theory. The anti-realist position I have developed here is built on a conscious awareness of the fact that there is little reason to believe that our contemporary theories are immune from being replaced in the future. That is, despite their success and the fact that they are the culmination of a sequence of successively better theories, no one has identified a feature of our current theories that distinguishes them from the theories they replaced in a principled manner that would warrant an inference to their truth or approximate truth. That, though, is the sort of challenge that realists must address if they are to alleviate the legitimate skeptical worries of anti-realists. Indeed, I have argued that one of the chief lessons we can learn from the history of science is that scientists of earlier periods probably thought much the same about their successful theories as contemporary scientists feel today about their theories. But until some distinguishing feature is identified that separates contemporary from past successful theories, a feature that earlier generations could not appeal to in an effort to separate their successful theories from the theories of their predecessors, the realists' optimism about today's successful theories is not warranted.

Some contemporary realists have acknowledged the fact of theory change but insist that the continuity through theory change, at least at some level, provides grounds for realism. Structural realists, for example, place a great deal of significance on the formulas that have persisted through changes of theory, Fresnel's equations being a frequently discussed example (see, for example, Worrall 1989). Other realists emphasize continuity in other respects (see, for example, Psillos 1999). Despite their differences, it seems that most realists put great stock in inferences to the best explanation. That is, they abductively infer that the best explanation for the remarkable progress and continuity in science is the truth or approximate truth of aspects of our theories. These realists, though, have had to help themselves to a pattern of inference that is itself an object of criticism in the debate. For this reason, the realists' optimism is unwarranted.

I am not suggesting that all inferences to the best explanation are problematic. As many have noted, inference to the best explanation is a pattern of reasoning that plays an important role in both everyday life and scientific inquiry. But I have argued that the conditions under which such inferences are made in the context of the realism/anti-realism debate are not the sorts of conditions where they are apt to be warranted. Most significantly, the realist often wants to make an inference about the truth or approximate truth of aspects of a theory when in fact scientists have considered only a narrow range of hypotheses or explanations. Again, it is here where we learn something from the history of science. Inferences to the best explanation made on the basis of a consideration of relatively few alternatives about speculative matters have frequently been shown to be unwarranted. The anti-realist is thus struck by the excessive confidence of the realist, by the apparent assumption that we have finally transcended the sorts of difficulties that have always proved so challenging to our predecessors. There is little evidence that we have finally transcended those sorts of challenging difficulties. Thus, I argue that a cautious attitude about claims of theoretical knowledge is warranted.

In addition to relying on inferences to the best explanation, realists have traditionally appealed to the theoretical virtues – simplicity, breadth of scope, and the like – to support their realism. But I have argued that these appeals to the theoretical virtues do not warrant the sorts of inferences that realists seek to draw. That is, rather than supporting an inference to the truth or approximate truth of a theory, an appeal to the theoretical virtues can only support an inference to the relative superiority of one theory over another.

I have also provided an account of how our scientific theories can be as successful as they are even if they may misrepresent aspects of the underlying unobservable reality they purport to model. It is no miracle that our theories can yield empirical successes, and do so on an ongoing basis. Scientists discard theories if they are not empirically successful. Leaving aside the issue of whether our theories are false or approximately true, I have argued that many of them are apt to be discarded in the future, when scientists extend their application to domains they were not originally designed to model. This is one significant reason why I have insisted that radical theory change is apt to continue to play a role in the future development of science.

It seems that the time is ripe to reconsider the viability of anti-realism. Though none of the arguments I present in the book considered alone can provide a conclusive argument against realism, collectively, the various strands of argumentation I present should persuade readers that realists face significant challenges and that anti-realism is a more viable position than commonly thought.

References

Alai, M. 2014. "Novel Predictions and the No Miracles Argument," *Erkenntnis*, 79, 297–326.

Alvarez, S., J. Sales, and M. Seco. 2008. "On Books and Chemical Elements," *Foundations of Chemistry*, 10, 79–100.

Barker, P. 2001. "Incommensurability and Conceptual Change during the Copernican Revolution," in P. Hoyningen-Huene and H. Sankey (eds.), *Incommensurability and Related Matters*. Dordrecht: Springer, pages 241–273.

Barker, P., and B. R. Goldstein. 1998. "Realism and Instrumentalism in Sixteenth Century Astronomy: A Reappraisal," *Perspectives on Science*, 6: 3, 232–258.

Barnes, B., and D. Bloor. 1982. "Relativism, Rationalism and the Sociology of Knowledge," in M. Hollis and S. Lukes (eds.), *Rationality and Relativism*. Cambridge, MA: MIT Press, pages 21–47.

Ben David, J. 1971. *The Scientist's Role in Society: A Comparative Study*. Englewood Cliffs, NJ: Prentice-Hall.

Biddle, J. 2013. "State of the Field: Transient Underdetermination and Values in Science," *Studies in History and Philosophy of Science*, 44: 1, 124–133.

Bird, A. 2000. *Thomas Kuhn*. Princeton, NJ: Princeton University Press.

Bishop, M. 2003. "The Pessimistic Induction, the Flight to Reference and the Metaphysical Zoo," *International Studies in the Philosophy of Science*, 17: 2, 161–178.

Blackburn, S. 2005. *Truth: A Guide*. Oxford: Oxford University Press.

Boyd, R. N. 1980. "Scientific Realism and Naturalistic Epistemology," *PSA: Proceedings of the Biennial Philosophy of Science Association*, Vol. 2: Symposia and Invited Papers (1980), 613–662.

1983. "On the Current Status of the Issue of Scientific Realism," *Erkenntnis*, 19: 1–3, 45–90.

1985. "*Lex orandi est lex credendi*," in P. M. Churchland and C. A. Hooker (eds.), *Images of Science: Essays on Realism and Empiricism, with a Reply from Bas C. van Fraassen*. Chicago: University of Chicago Press, pages 3–34.

Brahe, T. 1598/1946. *Tycho Brahe's Description of His Instruments and Scientific Work as given in Astronomiae Instauratae Mechanica*, translated and edited by H. Ræder, E. Strömgren, and B. Strömgren. København: Det Kongelige Danske Videnskabernes Selskab.

1588/1970. *On the Most Recent Phoenomena of the Aetherial World*, excerpted in M. B. Hall (ed.), *Nature and Nature's Law: Documents of the Scientific Revolution*. New York: Harper and Row, pages 58–66.

Brown, J. R. 1985. "Explaining the Success of Science," *Ratio*, XXVII: 1, 49–66.

Brush, S. G. 1994. "Dynamics of Theory Change: The Role of Predictions," *PSA: Proceedings of the Biennial Meeting of the Philosophy of Science Association*, Vol. 1994, Vol. 2: Symposia and Invited Papers, 133–145.

Campbell, D. T. 1977. "Comments on 'The Natural Selection Model of Conceptual Evolution,'" *Philosophy of Science*, 44, 502–507.

Carlson, A. K. 1996. "Lead Isotope Analysis of Human Bone for Addressing Cultural Affinity: A Case Study from Rocky Mountain House, Alberta," *Journal of Archaeological Science*, 23, 557–567.

Carnap, R. 1950. "Empiricism, Semantics, and Ontology," *Revue Internationale de Philosophie*, 4, 20–40.

Carrier, M. 1991. "What Is Wrong with the Miracle Argument?," *Studies in History and Philosophy of Science*, 22: 1, 23–36.

Cartwright, N. 1983. *How the Laws of Physics Lie*. Oxford: Clarendon Press.

Cedarbaum, D. G. 1983. "Paradigms," *Studies in History and Philosophy of Science*, 14: 3, 173–213.

Chakravartty, A. 2008. "What You Don't Know Can't Hurt You: Realism and the Unconceived," *Philosophical Studies*, 137, 149–158.

2007. *A Metaphysics for Scientific Realism: Knowing the Unobservable*. Cambridge: Cambridge University Press.

Chalmers, A. F. 2013. *What Is This Thing Called Science?*, 4th edition. Indianapolis: Hackett Publishing.

Chang, H. 2012. *Is Water H_2O?: Evidence, Realism and Pluralism*. Dordrecht: Springer.

Cherniak, C. 1986. *Minimal Rationality*. Cambridge, MA: MIT Press.

Christianson, J. R. 2000. *On Tycho's Island: Tycho Brahe and His Assistants, 1570–1601*. Cambridge: Cambridge University Press.

Copernicus, N. 1543/1995. *On the Revolutions of Heavenly Spheres*, translated by C. G. Wallis. Amherst, NY: Prometheus Books.

Creath, R. 2007. "Vienna, the City of Quine's Dreams," in A. Richardson and T. Uebel (eds.), *The Cambridge Companion to Logical Empiricism*. Cambridge: Cambridge University Press, pages 332–345.

Cutnell, J. D., and K. W. Johnson. 2001. *Physics*, 5th edition. New York: John Wiley & Sons.

Darwin, C. 1859/2003. *On the Origin of Species by Means of Natural Selection*, edited by J. Carroll. Peterborough, ON: Broadview Press, Ltd.

Dear, P. 2001. *Revolutionizing the Sciences: European Knowledge and Its Ambitions, 1500–1700*. Princeton, NJ: Princeton University Press.

Dellsén, F. Forthcoming. "Realism and the Absence of Rivals," *Synthese*. DOI: 10.1007/s11229–016–1059–3

Devitt, M. 2014. "Realism/Anti-realism," in M. Curd and S. Psillos (eds.), *The Routledge Companion to Philosophy of Science*. London: Routledge, pages 256–267.

2011. "Are Unconceived Alternatives a Problem for Scientific Realism?," *Journal for General Philosophy of Science*, 42: 2, 285–293.

Di Bono, M. 1995. "Copernicus, Amico, Fracastoro and the Tūsī Device: Observations on the Use and Transmission of a Model," *Journal for the History of Astronomy*, 26, 133–154.

Dicken, P. 2016. *A Critical Introduction to Scientific Realism*. London: Bloomsbury Academic.

Donovan, A., L. Laudan, and R. Laudan (eds.). 1988. *Scrutinizing Science: Empirical Studies of Scientific Change*. Synthese Library (Studies in Epistemology, Logic, Methodology, and Philosophy of Science), Vol. 193. Dordrecht: Springer.

Doppelt, G. 2005. "Empirical Success or Explanatory Success: What Does Current Scientific Realism Need to Explain?," *Philosophy of Science*, 72, 1076–1087.

Dreyer, J. L. E. 1906/1953. *A History of Astronomy from Thales to Kepler*, 2nd edition. New York: Dover Publications.

Duhem, P. 1908/1969. *To Save the Phenomena: An Essay on the Concept of Physical Theory from Plato to Galileo*, translated by E. Doland and C. Mascher. Chicago: University of Chicago Press.

1906/1954. *The Aim and Structure of Physical Theory*, translated by P. P. Wiener. Princeton, NJ: Princeton University Press.

Earman, J. 1993. "Underdetermination, Realism, and Reason," in *Midwest Studies in Philosophy*, 18: 1, 19–38.

Edge, D. O. and M. J. Mulkay. 1976. *Astronomy Transformed: The Emergence of Radio Astronomy in Britain*. New York: John Wiley and Sons.

Evans, J. 1998. *The History and Practice of Ancient Astronomy*. Oxford: Oxford University Press.

Everts, S. 2010. "When Science Went International: Looking back 150 years at the conference that led to the assembly of the periodic table," *Chemical & Engineering News* (September 3, 2010). https://pubs.acs.org/cen/science/88/8836sci1.html (accessed December 28, 2017).

Fahrbach, L. 2017. "Scientific Revolutions and the Explosion of Scientific Evidence," *Synthese*, 194, 5039–5072.

　　2011. "How the Growth of Science Ends Theory Change," *Synthese*, 180: 2, 139–155.

Feyerabend, P. K. 1988. *Against Method*, revised edition. London: Verso.

Fine, A. 1984. "The Natural Ontological Attitude," in J. Leplin (ed.), *Scientific Realism*. Berkeley and Los Angeles: University of California Press, pages 83–107.

Fisch, M. 2017. *Creatively Undecided: Toward a History and Philosophy of Scientific Agency*. Chicago: University of Chicago Press.

Fodor, J., and E. Lepore. 1992. *Holism: A Shopper's Guide*. Oxford: Blackwell Publishers.

Forster, M., and E. Sober. 1994. "How to Tell When Simpler, More Unified, or Less *Ad Hoc* Theories Will Provide More Accurate Predictions," *British Journal for the Philosophy of Science*, 45: 1, 1–35.

Frost-Arnold, G. 2011. "From the Pessimistic Induction to Semantic Antirealism," *Philosophy of Science*, 78: 5, 1131–1142.

　　2010. "The No-Miracles Argument for Realism: Inference to an Unacceptable Explanation," *Philosophy of Science*, 77: 1, 35–58.

Gade, J. A. 1947. *The Life and Times of Tycho Brahe*. Princeton, NJ: Princeton University Press.

Galilei, G. 1632/2001. *Dialogue Concerning the Two Chief World Systems*, translated and with revised notes by S. Drake. New York: Modern Library.

　　1615/2008. *Letter to the Grand Duchess Christina*, in M. Finocchiaro (ed.), *The Essential Galileo*. Indianapolis: Hackett Publishing, pages 109–145.

　　1612/2010. "Galileo's First Letter on Sunspots," in E. Reeves and A. van Helden (eds.), *Galileo Galilei and Christoph Scheiner: On Sunspots*. Chicago: University of Chicago Press, pages 89–105.

　　1610/2008. *The Sidereal Messenger*, in M. Finocchiaro (ed.), *The Essential Galileo*. Indianapolis: Hackett Publishing, pages 45–84.

Galison, P. 1987. *How Experiments End*. Chicago: University of Chicago Press.

Gelfert, A. 2016. *How to Do Science with Models: A Philosophical Primer*. Dordrecht: Springer International Publishing.

Giere, R. N. 1988. *Explaining Science: A Cognitive Approach*. Chicago: University of Chicago Press.

Gillies, D. 1993. *Philosophy of Science in the Twentieth Century: Four Central Themes*. Oxford: Blackwell Publishers.

Gingerich, O. 2004. *The Book That Nobody Read: Chasing the Revolutions of Nicolaus Copernicus*. London: Penguin Books.

1978/1993. "Early Copernican Ephemerides," in O. Gingerich (ed.), *The Eye of Heaven: Ptolemy, Copernicus, Kepler*. New York: Academic Institute of Physics, pages 205–220.

1975a/1993. "Kepler's Place in Astronomy," in O. Gingerich (ed.), *The Eye of Heaven: Ptolemy, Copernicus, Kepler*. New York: Academic Institute of Physics, pages 331–347.

1975b/1993. "'Crisis' versus Aesthetic in the Copernican Revolution," in O. Gingerich (ed.), *The Eye of Heaven: Ptolemy, Copernicus, Kepler*. New York: American Institute of Physics, pages 193–204.

1974/1993. "The Astronomy and Cosmology of Copernicus," in O. Gingerich (ed.), *The Eye of Heaven: Ptolemy, Copernicus, Kepler*. New York: American Institute of Physics, pages 161–184.

1971/1993. "Mercury Theory from Antiquity to Kepler," in O. Gingerich (ed.), *The Eye of Heaven: Ptolemy, Copernicus, Kepler*. New York: American Institute of Physics, pages 379–387.

Goldstein, B. R. 1996. "The Pre-telescopic Treatment of the Phases and Apparent Size of Venus," *Journal for the History of Astronomy*, 27, 1–12.

Green, R. E., et al. 2010. "A Draft Sequence of the Neandertal Genome," *Science*, 328, 5979 (May 7, 2010), 710–722.

Hacking, I. 1999. *The Social Construction of What?* Cambridge, MA: Harvard University Press.

1983. *Representing and Intervening: Introductory Topics in the Philosophy of Natural Science*. Cambridge: Cambridge University Press.

Hardin, C. L., and A. Rosenberg. 1982. "In Defense of Convergent Realism," *Philosophy of Science*, 49, 604–615.

Harker, D. 2010. "Two Arguments for Scientific Realism Unified," *Studies in History and Philosophy of Science*, 41, 192–202.

Heilbron, J. L. 2005. "Noble Gases," in J. L. Heilbron (ed.), *The Oxford Guide to the History of Physics and Chemistry*. Oxford: Oxford University Press, pages 230–231.

Hempel, C. 1966. *Philosophy of Natural Science*. Upper Saddle River, NJ: Prentice Hall.

Hendry, R. F. 2012. "Chemical Substances and the Limits of Pluralism," *Foundations of Chemistry*, 14, 55–68.

Henry, J. 2012. *A Short History of Scientific Thought*. Houndsmill, Basingstoke: Palgrave Macmillan.

Hesse, M. 1976. "Truth and the Growth of Scientific Knowledge," *PSA: Proceedings of the Biennial Meeting of the Philosophy of Science Association*, 2, 261–280.

1963. "Review of Thomas S. Kuhn's *The Structure of Scientific Revolutions*," *Isis*, 54: 2, 286–287.

Hoskin, M. 1997a. "Astronomy in Antiquity," in M. Hoskin (ed.), *Cambridge Illustrated History of Astronomy*. Cambridge: Cambridge University Press, pages 22–47.

1997b. "From Geometry to Physics: Astronomy Transformed," in M. Hoskin (ed.), *Cambridge Illustrated History of Astronomy*. Cambridge: Cambridge University Press, pages 98–143.

Howson, C. 2000. *Hume's Problem: Induction and the Justification of Belief*. Oxford: Oxford University Press.

Hoyningen-Huene, P. 2008. "Thomas Kuhn and the Chemical Revolution," *Foundations of Chemistry*, 10: 2, 101–115.

Hudson, J. 1992. *The History of Chemistry*. New York: Chapman and Hall.

Hull, D. L. 2001. "Why Scientists Behave Scientifically," in D. L. Hull (ed.), *Science and Selection: Essays on Biological Evolution and the Philosophy of Science*. Cambridge: Cambridge University Press, pages 135–138.

1988. *Science as a Process: An Evolutionary Account of the Social and Conceptual Development of Science*. Chicago: University of Chicago Press.

Idhe, A. J. 1961. "The Karlsruhe Congress: A Centennial Retrospect," *Journal of Chemical Education*, 38: 2, 83–86.

Inquisition. 1633/2008. "Inquisition's Sentence (22 June 1633)," in M. A. Finocchiaro (ed.), *The Essential Galileo*. Indianapolis: Hackett Publishing, pages 288–293.

Ivanona, M. 2015. "Conventionalism about What?: Where Duhem and Poincaré Part Ways," *Studies in History and Philosophy of Science*, 54, 80–89.

James, W. 1907/1949. "Pragmatism: A New Name for Some Old Ways of Thinking", in W. James (ed.), *Pragmatism: A New Name for Some Old Ways of Thinking together with Four Related Essays Selected from The Meaning of Truth*. New York: Longmans, Green and Company.

Kaji, M. 2002. "D. I. Mendeleev's Concept of Chemical Elements and *The Principles of Chemistry*," *Bulletin for the History of Chemistry*, 27: 1, 4–16.

Keas, M. N. Forthcoming. "Systematizing the Theoretical Virtues," *Synthese*. DOI: 10.1007/s11229–017–1355–6

Kepler, J. 1618–1621/1995. "Epitome of Copernican Astronomy", in J. Kepler (ed.), *Epitome of Copernican Astronomy and Harmonies of the World*, translated by C. G. Wallis. Amherst, NY: Prometheus Books.

Kitcher, P. 1993. *Advancement of Science: Science without Legend, Objectivity without Illusions*. Oxford: Oxford University Press.

Koestler, A. 1959/1964. *Sleepwalkers: A History of Man's Changing Vision of the Universe*. London: Penguin Books.

Kuhn, T. S. 2000. "A Discussion with Thomas S. Kuhn," in J. Conant and J. Haugeland (eds.), *The Road since Structure: Philosophical Essays, 1970–1993, with an Autobiographical Interview*. Chicago: University of Chicago Press, pages 255–323.

1992/2000. "The Trouble with the Historical Philosophy of Science," in J. Conant and J. Haugeland (eds.), *The Road since Structure: Philosophical Essays, 1970–1993, with an Autobiographical Interview*. Chicago: University of Chicago Press, pages 105–120.

1991/2000. "The Road since *Structure*," in T. S. Kuhn's J. Conant and J. Haugeland (eds.), *The Road since Structure: Philosophical essays, 1970–1993, with an Autobiographical Interview*. Chicago: University of Chicago Press, pages 90–104.

1987/2000. "What Are Scientific Revolutions?," in J. Conant and J. Haugeland (eds.), *The Road since Structure: Philosophical Essays, 1970–1993, with an Autobiographical Interview*. Chicago: University of Chicago Press, pages 13–32.

1977. "Objectivity, Value Judgment, and Theory Choice," in T. S. Kuhn (ed.), *The Essential Tension: Selected Studies in Scientific Tradition and Change*. Chicago: University of Chicago Press, pages 320–339.

1977b. "Preface," in *The Essential Tension: Selected Studies in Scientific Tradition and Change*. Chicago: University of Chicago Press, pages ix–xxiii.

1976/1977. "The Relations between the History and the Philosophy of Science," in T. S. Kuhn (ed.), *The Essential Tension: Selected Studies in Scientific Tradition and Change*. Chicago: University of Chicago Press, pages 3–20.

1968/1977. "The History and the Philosophy of Science," in T. S. Kuhn (ed.), *Essential Tension: Selected Studies in Scientific Tradition and Change*. Chicago: University of Chicago Press, pages 3–20.

1962/2012. *Structure of Scientific Revolutions*, 4th edition, with an introductory essay by Ian Hacking. Chicago: University of Chicago Press.

1957. *The Copernican Revolution: Planetary Astronomy in the Development of Western Thought*. Cambridge, MA: Harvard University Press.

Kukla, A. 1996a. "Antirealist Explanations of the Success of Science," *Philosophy of Science*, 63 (Proceedings), S298–S305.

1996b. "Does Every Theory Have Empirically Equivalent Rivals?," *Erkenntnis*, 44: 2, 137–166.

Kusch, M. 2015. "Scientific Pluralism and the Chemical Revolution," *Studies in History and Philosophy of Science*, 49, 69–79.

Ladyman, J. 2002. *Understanding Philosophy of Science*. London: Routledge.

Lange, M. 2002. "Baseball, Pessimistic Inductions, and the Turnover Fallacy," *Analysis*, 62: 4, 281–285.

Lattis, J. M. 1994. *Between Copernicus and Galileo: Christoph Clavius and the Collapse of Ptolemaic Cosmology*. Chicago: University of Chicago Press.

Laudan, L. 2004. "The Epistemic, the Cognitive, and the Social," in P. Machamer and G. Wolters (eds.), *Science, Values, and Objectivity*. Pittsburgh: University of Pittsburgh Press, pages 14–23.

1990. "Demystifying Underdetermination," in C. W. Savage (ed.), *Scientific Theories*, Vol. 14, *Minnesota Studies in the Philosophy of Science*. Minneapolis: University of Minnesota Press, pages 267–297.

1984a. *Science and Values: The Aims of Science and Their Role in Scientific Debate*. Berkeley: University of California Press.

1984b. "Explaining the Success of Science," in in J. T. Cushing, C. F. Delaney, and G. Gutting (eds.), *Science and Reality: Recent Work in the Philosophy of Science*. Notre Dame, IN: University of Notre Dame, 83–105.

1981. "A Confutation of Convergent Realism," *Philosophy of Science*, 48: 19–49.

1977. *Progress and Its Problems*. Berkeley and Los Angeles: University of California Press.

Laudan, L, and J. Leplin. 1991. "Empirical Equivalence and Underdetermination," *Journal of Philosophy*, 88: 9, 449–472.

Law, J. 1976. "The Development of Specialties in Science: The Case of X-Ray Protein Crystallography," in G. Lemaine, R. MacLeod, M. Mulkay, and P. Weingart (eds.), *Perspectives on the Emergence of Scientific Disciplines*. Chicago: Aldine Publishing, pages 1–23.

Leplin, J. 1997. *A Novel Defense of Scientific Realism*. Oxford: Oxford University Press.

Lewis, P. 2001. "Why the Pessimistic Induction Is a Fallacy," *Synthese*, 129: 3, 371–380.

Lindberg, D. C. 2007. *The Beginnings of Western Science: The European Scientific Tradition in Philosophical, Religious, and Institutional Context, Prehistory to A.D. 1450*, 2nd Edition. Chicago: University of Chicago Press.

Lipton, P. 2004. *Inference to the Best Explanation*, 2nd Edition. London: Routledge.

1993/1996. "Is the Best Good Enough?," in D. Papineau (ed.), *The Philosophy of Science*. Oxford: Oxford University Press, pages 93–106.

Longino, H. E. 2001. *The Fate of Knowledge*. Princeton, NJ: Princeton University Press.

1995. "Gender, Politics, and the Theoretical Virtues," *Synthese*, 104, 383–397.

1990. *Science as Social Knowledge: Values and Objectivity in Scientific Inquiry.* Princeton, NJ: Princeton University Press.

Lyons, T. D. Forthcoming. "Epistemic Selectivity, Historical Threats, and the Non-epistemic Tenets of Scientific Realism," *Synthese.* DOI:10.1007/s11229-016-1103-3

2013. "A Historically Informed *Modus Ponens* against Scientific Realism: Articulation, Critique, and Restoration," *International Studies in the Philosophy of Science*, 27: 4, 369–392.

2012. "Axiological Scientific Realism and Methodological Prescription," in H. W. Regt, S. Hartmann, and S. Okasha (eds.), *EPSA Philosophy of Science: Amsterdam 2009. The European Philosophy of Science Association Proceedings*, Vol. 1. Dordrecht: Springer.

2006. "Scientific Realism and the Strategema de Divide et Impera," *British Journal for the Philosophy of Science*, 57, 537–560.

2002. "Scientific Realism and the Pessimistic Meta-Modus Tollens," in S. Clarke and T. D. Lyons (eds.), *Recent Themes in the Philosophy of Science.* Dordrecht: Kluwer Academic Publishers, pages 63–90.

Mach, E. 1911. *History and Root of the Principle of the Conservation of Energy*, translated and annotated by P. E. B. Jourdain. Chicago: Open Court.

1897/1984. *The Analysis of Sensations and the Relation of the Physical to the Psychical*, translated by C. M. Williams. La Salle, IL: Open Court.

1892. "Facts and Mental Symbols," *The Monist*, 2: 2, 198–208.

Magnus, P. D. 2006. "What's New About the New Induction?," *Synthese*, 148, 295–301.

Magnus, P. D., and C. Callender. 2004. "Realist Ennui and the Base Rate Fallacy," *Philosophy of Science*, 71: 3, 320–338.

Masterman, M. 1970. "The Nature of a Paradigm," in I. Lakatos and A. Musgrave (eds.), *Criticism and the Growth of Knowledge: Proceedings of the International Colloquium in the Philosophy of Science, London, 1965*, Vol. 4. Cambridge: Cambridge University Press, pages 59–89.

McMullin, E. 2008. "The Virtues of a Good Theory," in S. Psillos and M. Curd (eds.), *The Routledge Companion to Philosophy of Science.* London: Routledge, pages 498–508.

1993. "Rationality and Paradigm Change in Science," in P. Horwich (ed.), *World Changes: Thomas Kuhn and the Nature of Science.* Cambridge, MA: MIT Press, pages 55–78.

Merton, R. K. 1973. *The Sociology of Science: Theoretical and Empirical Investigations*, edited by N. W. Storer. Chicago: University of Chicago Press.

Mizrahi, M. 2016. "The History of Science as a Graveyard of Theories: A Philosophers' Myth," *International Studies in the Philosophy of Science*, 30: 3, 263–278.

2013. "The Pessimistic Induction: A Bad Argument Gone too Far," *Synthese*, 190: 15, 3209–3226.

Montaigne, M. 1580/1948. "Apology for Raymond Sebond," in *The Complete Essays of Montaigne*, translated by D. M. Frame. Stanford, CA: Stanford University Press, pages 318–457.

Musgrave, A. 1988. "The Ultimate Argument for Scientific Realism," in R. Nola (ed.), *Relativism and Realism in Science*. Dordrecht: Kluwer Academic Publishers, pages 229–252.

Newton-Smith, W. H. 2000. "Underdetermination of Theory by Data," in W. H. Newton-Smith (ed.), *A Companion to the Philosophy of Science*. Malden, MA: Blackwell Publishers, pages 532–536.

Niiniluoto, I. 1999. *Critical Scientific Realism*. Oxford: Oxford University Press.

Nola, R. 2008. "The Optimistic Meta-induction and Ontological Continuity: The Case of the Electron," in L. Soler, H. Sankey, and P. Hoyningen-Huene (eds.), *Rethinking Scientific Change and Theory Comparison: Stabilities, Ruptures, Incommensurabilities*. Dordrecht: Springer, pages 159–202.

Osiander, A. 1543/1995. "Introduction: To the Reader Concerning the Hypotheses of this Work," in N. Copernicus, *On the Revolutions of the Heavenly Spheres*, translated by C. G. Wallis. Amherst, NY: Prometheus Books, pages 3–4.

Papineau, D. 1996. "Introduction," in D. Papineau (ed.), *The Philosophy of Science*. Oxford: Oxford University Press, pages 1–20.

Poincaré, H. 1905/2001. "Science and Hypothesis," in *The Value of Science: Essential Writings of Henri Poincaré*. New York: Dover Publications, pages 1–178.

1913/2001. "The Value of Science," in *The Value of Science: Essential Writings of Henri Poincaré*. New York: Dover Publications, pages 181–353.

Popper, K. R. 1971. *Objective Knowledge: An Evolutionary Approach*. Oxford: Clarendon Press.

1970. "Normal Science and Its Dangers," in I. Lakatos and A. Musgrave (eds.), *Criticism and the Growth of Knowledge: Proceedings of the International Colloquium in the Philosophy of Science, London 1965*, Vol. 4, reprinted with corrections (1972). Cambridge: Cambridge University Press, pages 51–58.

1963a. *Conjectures and Refutations: The Growth of Scientific Knowledge*. London: Routledge.

1963b/1963a. "Truth, Rationality, and the Growth of Knowledge," in K. R. Popper (1963a) (ed.), *Conjectures and Refutations: The Growth of Scientific Knowledge*. London: Routledge, pages 291–338.

1957/1963a. "Science: Conjectures and Refutations," in K. R. Popper (1963a) (ed.), *Conjectures and Refutations: The Growth of Scientific Knowledge*. London: Routledge, pages 43–86.

1956/1963a. "Three Views Concerning Human Knowledge," in K. R. Popper (1963a) (ed.), *Conjectures and Refutations: The Growth of Scientific Knowledge*. London: Routledge, pages 130–160.

1952/1963a. "The Nature of Philosophical Problems and their Roots in Science," in K. R. Popper (1963a) (ed.), *Conjectures and Refutations: The Growth of Scientific Knowledge*. London: Routledge, pages 88–129.

1935/2002. *The Logic of Scientific Discovery*, English edition. London: Routledge.

Price, D. de Solla. 1963. *Little Science, Big Science*. New York: Columbia University Press.

Psillos, S. 1999. *Scientific Realism: How Science Tracks Truth*. London: Routledge.

1996. "Scientific Realism and the 'Pessimistic Induction,'" *Philosophy of Science*, 63 (Proceedings), S306–S314.

Ptolemy, C. 1952. *The Almagest*. London: Encyclopedia Britannica.

Putnam, H. 1978. *Meaning and the Moral Sciences*. London: Routledge & Kegan Paul.

1975. *Mathematics, Matter and Method: Philosophical Papers*, Vol. 1. Cambridge: Cambridge University Press.

Pyle, A. 2000. "The Rationality of the Chemical Revolution," in R. Nola and H. Sankey (eds.), *After Popper, Kuhn, and Feyerabend*. Dordrecht: Kluwer Academic Publishers, pages 99–124.

Quine, W. V. 1969. "Epistemology Naturalized," in W. V. Quine (ed.), *Ontological Relativity and Other Essays*. New York: Columbia University Press, pages 69–90.

1951/1953. "Two Dogmas of Empiricism," in W. V. Quine (ed.), *From a Logical Point of View*. Cambridge, MA: Harvard University Press, pages 20–46.

1951. "Two Dogmas of Empiricism," *Philosophical Review*, 60: 1, 20–43.

Reeves, E., and A. van Helden. 2010. *On Sunspots: Galileo Galilei and Christoph Scheiner*. Chicago: University of Chicago Press.

Rescher, N. 1978. *Scientific Progress: A Philosophical Essay on the Economics of Research in Natural Science*. Pittsburgh: University of Pittsburgh Press.

1987. *Scientific Realism: A Critical Reappraisal*. Dordrecht: D. Reidel Publishing.

Restrepo, G., and L. Pachón. 2007. "Mathematical Aspects of the Periodic Law," *Foundations of Chemistry*, 9, 189–214.

Rohland, N., D. Reich, S. Mallick, M. Meyer, R. E. Green, N. J. Georgiadis, A. L. Roca, and M. Hofreiter. 2010. "Genomic DNA Sequences from

Mastodon and Woolly Mammoth Reveal Deep Speciation of Forest and Savannah Elephants," *PLOS Biology* (December 21, 2010).

Rolin, K. 2006. "The Bias Paradox in Feminist Standpoint Epistemology," *Episteme*, 3: 1–2, 125–136.

Rosen, G. 1994. "What Is Constructive Empiricism?," *Philosophical Studies*, 74: 2, 143–178.

Rosen, E. 1939/1959. "Introduction," in E. Rosen (ed.), *Three Copernican Treatises*, 2nd edition, translated with introduction and notes by E. Rosen. Mineola, NY: Dover Publications, pages 3–53.

Roush, S. 2010. "Optimism about the Pessimistic Induction," in P. D. Magnus and J. Busch (eds.), *New Waves in Philosophy of Science*. Houndsmill, UK: Palgrave Macmillan, pages 29–58.

Rowbottom, D. P. 2014. "Aimless Science," *Synthese*, 191, 1211–1221.

Ruhmkorff, S. 2013. "Global and Local Pessimistic Meta-Inductions," *International Studies in the Philosophy of Science*, 27: 4, 409–428.

Saatsi, J. T. 2005. "On the Pessimistic Induction and Two Fallacies," *Philosophy of Science*, 72, 1088–1098.

Sankararaman, S., N. Patterson, H. Li, S. Pääbo, and D. Reich. 2012. "The Date of Interbreeding Between Neandertals and Modern Humans," *PLOS Genetics*, 8: 10, e1002947. DOI:10.1371/journal.pgen.1002947

Scerri, E. R. 2016. *A Tale of Seven Scientists and a New Philosophy of Science*. Oxford: Oxford University Press.

2013. *A Tale of Seven Elements*. Oxford: Oxford University Press.

2012. "A Critique of Weisberg's View on the Periodic Table and Some Speculations on the Nature of Classifications," *Foundations of Chemistry*, 14: 3, 275–284.

2011. *The Periodic Table: A Very Short Introduction*. Oxford: Oxford University Press.

2007. *The Periodic Table: Its Story and Significance*. Oxford: Oxford University Press.

Scerri, E. R., and J. Worrall. 2001. "Prediction and the Periodic Table," *Studies in History and Philosophy of Science*, 32: 3, 407–452.

Scheffler, I. 1967. *Science and Subjectivity*. Indianapolis: Bobbs-Merrill.

Shank, M. H. 2002. "Regiomontanus on Ptolemy, Physical Orbs, and Astronomical Fictionalism: Goldsteinian Themes in the 'Defense of Theon against George of Trebizond,'" *Perspectives on Science*, 10: 2, 179–207.

Shapere, D. 1964/1980. "Review of the Structure of Scientific Revolutions," in G. Gutting (ed.), *Paradigms and Revolutions: Applications and Appraisals of Thomas Kuhn's Philosophy of Science*. Notre Dame, IN: University of Notre Dame Press, pages 27–38.

Shea, W. R. 1998. "Galileo's Copernicanism: The Science and the Rhetoric," in P. Machamer (ed.), *The Cambridge Companion to Galileo*. Cambridge: Cambridge University Press, pages 211–243.

Shea, W. R., and M. Artigas. 2003. *Galileo in Rome: The Rise and Fall of a Troublesome Genius*. Oxford: Oxford University Press.

Sklar, L. 1975. "Methodological Conservativism," *Philosophical Review*, 84: 3, 374–400.

Smart, J. J. C. 1963/2009. *Philosophy and Scientific Realism*. London: Routledge & Kegan Paul.

Soddy, F. 1913. "Intra-atomic Charge," *Nature*, 2301: 92 (December 4, 1913), 399–400.

Special Commission. 1632/2008. "Special Commission's Report on the *Dialogue* (September 1632)," in M. A. Finocchiaro (ed.), *The Essential Galileo*. Indianapolis: Hackett Publishing, pages 272–276.

Stanford, P. K. 2006. *Exceeding Our Grasp: Science, History, and the Problem of Unconceived Alternatives*. Oxford: Oxford University Press.

2003. "Pyrrhic Victories for Scientific Realism," *Journal of Philosophy*, C: 11, 553–572.

2001. "Refusing the Devil's Bargain: What Kind of Underdetermination Should We Take Seriously?," *Philosophy of Science*, 68: 3 (Proceedings), S1–S12.

Steel, D. 2010. "The Epistemic Values and the Argument from Inductive Risk," *Philosophy of Science*, 77, 14–34.

Suárez, M. 2009. "Fictions in Scientific Practice," in M. Suárez (ed.), *Fictions in Science: Philosophical Essays on Modeling and Idealization*. London: Routledge, pages 3–15.

Swerdlow, N. M. 2004. "An Essay on Thomas Kuhn's First Scientific Revolution, *The Copernican Revolution*," in *Proceedings of the American Philosophical Society*, 148: 1, 64–120.

1998. "Galileo's Discoveries with the Telescope and Their Evidence for the Copernican Theory," P. Machamer (ed.), *The Cambridge Companion to Galileo*. Cambridge: Cambridge University Press, pages 244–270.

Thagard, P. 1999. *How Scientists Explain Disease*. Princeton, NJ: Princeton University Press.

1990. "The Conceptual Structure of the Chemical Revolution," *Philosophy of Science*, 57, 183–209.

Thoren, V. E. 1967. "An Early Instance of Deductive Discovery: Tycho Brahe's Lunar Theory," *Isis*, 58: 1, 19–36.

1990. *The Lord of Uraniborg: A Biography of Tycho Brahe*, with contributions by J. R. Christianson. Cambridge: Cambridge University Press.

Thornton, B. F., and S. C. Burdette. 2010. "Finding Eka-iodine: Discovery Priority in Modern Times," *Bulletin for the History of Chemistry*, 35: 2, 86–96.

Tichý, P. 1974. "On Popper's Definition of Verisimilitude," *British Journal for the Philosophy of Science*, 25: 2, 155–160.

Toulmin, S. 1981. "Evolution, Adaptation, and Human Understanding," in M. B. Brewer and B. E. Collins (eds.), *Scientific Inquiry and the Social Sciences*. San Francisco: Jossey-Bass, pages 18–36.

Tredwell, K. A., and P. Barker. 2004. "Copernicus' First Friends: Physical Copernicanism from 1543 to 1610," *Filozofski Vestnik*, XXV: 2, 143–166.

Treiman, A. H., J. D. Gleason, and D. D. Bogard. 2000. "The SNC Meteorites Are from Mars," *Planetary and Space Science*, 48: 12–14, 1213–1230.

Trout, J. D. 2016. *Wondrous Truths: The Improbable Triumph of Modern Science*. Oxford: Oxford University Press.

Van Fraassen, B. C. 2007. "From a View of Science to a New Empiricism," in B. Monton (ed.), *Images of Empiricism: Essays on Science and Stances, with a Reply from Bas C. van Fraassen*. Oxford: Oxford University Press, pages 337–383.

1994. "Gideon Rosen on Constructive Empiricism," *Philosophical Studies*, 74: 2, 179–192.

1989. *Laws and Symmetry*. Oxford: Clarendon Press.

1980. *The Scientific Image*. Oxford: Clarendon Press.

Van Spronsen, J. W. 1969. *The Periodic System of Chemical Elements: A History of the First Hundred Years*. Amsterdam: Elsevier.

Vernot, B., and J. M. Akey. 2014. "Resurrecting Surviving Neandertal Lineages from Modern Human Genomes," *Science*, 343: 6174 (February 28, 2014), 1017–1021.

Vickers, P. 2013. "A Confrontation of Convergent Realism," *Philosophy of Science*, 80, 189–211.

Westman, R. S. 2011. *The Copernican Question: Prognostication, Skepticism, and Celestial Order*. Berkeley and Los Angeles: University of California Press.

1986/2003. "The Copernicans and the Churches," in M. Hellyer (ed.), *The Scientific Revolution*. Oxford: Blackwell Publishing, pages 46–71.

1975. "The Melanchthon Circle, Rheticus, and the Wittenberg Interpretation of the Copernican Theory," *Isis*, 66: 2, 164–193.

Worrall, J. 2012. "Miracles and Structural Realism," in E. Landry and D. Rickles (eds.), *Structural Realism: Structure, Object, and Causality*. Dordrecht: Springer, pages 77–95.

2007. "Miracles and Models: Why Reports of the Death of Structural Realism May Be Exaggerated," in A. O'Hear (ed.), *Philosophy of Science*, Supplement to *Philosophy*, Royal Institute of Philosophy Supplement: 61. Cambridge: Cambridge University Press, pages 125–154.

1989. "Structural Realism: The Best of Both Worlds?," *Dialectica*, 43: 1–2, 99–124.

Wray, K. B. 2017. "Kuhn's Influence on the Social Sciences," in A. Rosenberg and L. McIntyre (eds.), *Routledge Companion to Philosophy of Social Science*. New York: Routledge, pages 65–75.

  2015. "Pessimistic Inductions: Four Varieties," *International Studies in the Philosophy of Science*, 29: 1, 61–73.

  2015b. "The Methodological Defense of Realism Scrutinized," *Studies in History and Philosophy of Science*, 54, 74–79.

  2013. "Success and Truth in the Realism/Anti-realism Debate," *Synthese*, 190: 9, 1719–1729.

  2011. *Kuhn's Evolutionary Social Epistemology*. Cambridge: Cambridge University Press.

  2007. "Who Has Scientific Knowledge?," *Social Epistemology*, 21: 3, 337–347.

Wright, J. 2013. *Explaining Science's Success: Understanding How Scientific Knowledge Works*. Durham: Acumen.

Wrightsman, B. 1975. "Andreas Osiander's Contribution to the Copernican Achievement," in R. S. Westman (ed.), *The Copernican Achievement*. Berkeley and Los Angeles: University of California Press, pages 213–243.

Index

Printed in the United States
By Bookmasters